Planetary Protection Classification of Sample Return Missions from the Martian Moons

Committee on Planetary Protection Requirements for
Sample Return Missions from Martian Moons

Space Studies Board
Division on Engineering and Physical Sciences

and

European Space Sciences Committee
European Science Foundation
Strasbourg, France

A Consensus Study Report of

The National Academies of
SCIENCES · ENGINEERING · MEDICINE

EUROPEAN
SCIENCE
FOUNDATION

THE NATIONAL ACADEMIES PRESS
Washington, DC
www.nap.edu

THE NATIONAL ACADEMIES PRESS **500 Fifth Street, NW** **Washington, DC 20001**

This study is based on work supported by Contract NNH17CB02B between the National Academy of Sciences and the National Aeronautics and Space Administration and work supported by Contract 4000118481 between the European Science Foundation and the European Space Agency. Any opinions, findings, conclusions, or recommendations expressed in this publication do not necessarily reflect the views of any organization or agency that provided support for the project.

International Standard Book Number-13: 978-0-309-48859-4
International Standard Book Number-10: 0-309-48859-1
Digital Object Identifier: https://doi.org/10.17226/25357

Limited copies of this and other ESSC reports are available free of charge from:

European Science Foundation
Dr. Emmanouil Detsis
B.P. 90015
67080 Strasbourg Cedex
France
Phone: +33 388 767154 / essc@esf.org
http://www.essc.esf.org

Limited copies of this and other SSB reports are available free of charge from:

Space Studies Board
Keck Center of the National Academies of Sciences, Engineering, and Medicine
500 Fifth Street, NW, Washington, DC 20001
(202) 334-3477 / ssb@nas.edu
www.nationalacademies.org/ssb/ssb.html

Additional copies of this publication are available for sale from the National Academies Press, 500 Fifth Street, NW, Keck 360, Washington, DC 20001; (800) 624-6242 or (202) 334-3313; http://www.nap.edu.

Printed in the United States of America

Suggested citation: National Academies of Sciences, Engineering, and Medicine and the European Science Foundation. 2019. *Planetary Protection Classification of Sample Return Missions from the Martian Moons*. Washington, DC: The National Academies Press. doi: https://doi.org/10.17226/25357.

The National Academies of
SCIENCES · ENGINEERING · MEDICINE

The **National Academy of Sciences** was established in 1863 by an Act of Congress, signed by President Lincoln, as a private, nongovernmental institution to advise the nation on issues related to science and technology. Members are elected by their peers for outstanding contributions to research. Dr. Marcia McNutt is president.

The **National Academy of Engineering** was established in 1964 under the charter of the National Academy of Sciences to bring the practices of engineering to advising the nation. Members are elected by their peers for extraordinary contributions to engineering. Dr. C. D. Mote, Jr., is president.

The **National Academy of Medicine** (formerly the Institute of Medicine) was established in 1970 under the charter of the National Academy of Sciences to advise the nation on medical and health issues. Members are elected by their peers for distinguished contributions to medicine and health. Dr. Victor J. Dzau is president.

The three Academies work together as the **National Academies of Sciences, Engineering, and Medicine** to provide independent, objective analysis and advice to the nation and conduct other activities to solve complex problems and inform public policy decisions. The National Academies also encourage education and research, recognize outstanding contributions to knowledge, and increase public understanding in matters of science, engineering, and medicine.

Learn more about the National Academies of Sciences, Engineering, and Medicine at **www.nationalacademies.org**.

The National Academies of

SCIENCES · ENGINEERING · MEDICINE

Consensus Study Reports published by the National Academies of Sciences, Engineering, and Medicine document the evidence-based consensus on the study's statement of task by an authoring committee of experts. Reports typically include findings, conclusions, and recommendations based on information gathered by the committee and the committee's deliberations. Each report has been subjected to a rigorous and independent peer-review process and it represents the position of the National Academies on the statement of task.

Proceedings published by the National Academies of Sciences, Engineering, and Medicine chronicle the presentations and discussions at a workshop, symposium, or other event convened by the National Academies. The statements and opinions contained in proceedings are those of the participants and are not endorsed by other participants, the planning committee, or the National Academies.

For information about other products and activities of the National Academies, please visit www.nationalacademies.org/about/whatwedo.

SCIENCE CONNECT

EUROPEAN SCIENCE FOUNDATION

YOUR PARTNER IN SCIENCE

Based in Strasbourg, France, the **European Science Foundation (ESF)** was established in 1974 as an independent, nongovernmental, nonprofit organization to help its Member Organizations collaborate internationally on research programs. ESF helps meet this objective by providing administrative, management, and coordination services to independent scientific boards/committees and collaborative scientific projects, while harnessing in-depth knowledge of the European landscape and associated research communities.

Currently, ESF's focus lies in activities designed to support and sustain the funding and conduct of scientific research across Europe. This is a natural evolution of its traditional role, since the aim remains to promote scientific developments through collaborative actions, but with the emphasis shifting to helping research funding organizations carry out their decision-making processes.

ESF's aim is to serve and strengthen science by exploiting its assets to build and develop an organization that is self-financing and independent, but at the same time nonprofit and customer-service oriented. To this end it will be working alongside leading science funding institutions as well as national funding bodies.

The **European Space Sciences Committee (ESSC)** is an Expert Board of the ESF. The ESSC aims to underpin the role of space sciences and technology as pillars of the European space venture, support European visibility and enhance the position of Europe in global space initiatives, and assess the status and perspectives of European space activities on a regular basis.

Following the creation of the "Provisional Space Science Board for Europe" under the auspices of the UK Royal Society in 1974, the ESSC turned into a Standing Committee of the newly created ESF in 1975. The ESSC had grown out of the need for a collaborative effort that would ensure that European space scientists made their voices heard on the other side of the Atlantic, in an era when successive Apollo and space science missions had thrust the idea of space exploration into the collective conscious for the first time.

The ESSC's mission is to provide an independent voice on European space research and policy, and it remains just as relevant today as it acts as an interface with the European Space Agency (ESA), the European Commission, national space agencies, and ESF Member Organizations on space-related aspects.

COMMITTEE ON PLANETARY PROTECTION REQUIREMENTS FOR SAMPLE RETURN MISSIONS FROM MARTIAN MOONS

DAVID PEARCE, Northumbria University, United Kingdom, *Chair*
ANDRÉ ANTUNES, Edge Hill University, United Kingdom
ATHENA COUSTENIS, LESIA-Observatoire de Paris-Meudon, France
MICHAEL J. DALY, Uniformed Services University of the Health Sciences, United States
ABIGAIL A. FRAEMAN, Jet Propulsion Laboratory, United States
ANSGAR GRESHAKE, Museum für Naturkunde, Germany
GUY LIBOUREL, Observatoire de la Côte d'Azur, France
AKIKO NAKAMURA, Kobe University, Japan
FRANÇOIS POULET, Institute of Space Astrophysics, France
ROBIN PUTZAR, Fraunhofer Institute for High-Speed Dynamics, Germany
KALIAT T. RAMESH, Johns Hopkins University, United States
NORMAN H. SLEEP,[1] Stanford University, United States
SHINO SUZUKI, Japan Agency for Marine Earth Science and Technology, Japan
MEGAN BRUCK SYAL, Lawrence Livermore National Laboratory, United States
ERIN L. WALTON, MacEwan University, Canada

Staff

EMMANOUIL DETSIS, Science Officer, European Science Foundation
DAVID H. SMITH, Senior Staff Officer, Space Studies Board
MIA BROWN, Research Associate, Space Studies Board
ANDREA REBHOLZ, Program Coordinator, Space Studies Board
JONATHAN LUTZ, Lloyd V. Berkner Space Policy Intern, Space Studies Board

[1] Member, National Academy of Sciences.

EUROPEAN SPACE SCIENCES COMMITTEE

SPACE STUDIES BOARD

MARGARET KIVELSON, NAS, University of California, Los Angeles, *Chair*
JAMES H. CROCKER, NAE,[1] Lockheed Martin Space Systems Company (retired), *Vice Chair*
GREGORY P. ASNER, NAS, Carnegie Institution for Science
JEFF M. BINGHAM, Consultant
ADAM BURROWS, NAS, Princeton University
MARY LYNNE DITTMAR, Dittmar Associates
JEFF DOZIER, University of California, Santa Barbara
JOSEPH FULLER JR., Futron Corporation (retired)
SARAH GIBSON, National Center for Atmospheric Research
VICTORIA HAMILTON, Southwest Research Institute
CHRYSSA KOUVELIOTOU, NAS, George Washington University
DENNIS P. LETTENMAIER, NAE, University of California, Los Angeles
ROSALY M. LOPES, Jet Propulsion Laboratory
STEPHEN J. MACKWELL, American Institute of Physics
DAVID J. McCOMAS, Princeton University
LARRY PAXTON, Johns Hopkins University, Applied Physics Laboratory
ELIOT QUATAERT, University of California, Berkeley
BARBARA SHERWOOD LOLLAR, University of Toronto
HARLAN E. SPENCE, University of New Hampshire
MARK H. THIEMENS, NAS, University of California, San Diego
ERIKA WAGNER, Blue Origin
PAUL WOOSTER, Space Exploration Technologies
EDWARD L. WRIGHT, NAS, University of California, Los Angeles

Staff

MICHAEL H. MOLONEY, Director (through March 1, 2018)
RICHARD ROWBERG, Acting Director (March 1 through August 6, 2018)
COLLEEN HARTMAN, Director (beginning August 6, 2018)
CARMELA J. CHAMBERLAIN, Administrative Coordinator (through June 30, 2018)
TANJA PILZAK, Manager, Program Operations
CELESTE A. NAYLOR, Information Management Associate
MARGARET KNEMEYER, Financial Officer

[1] Member, National Academy of Engineering.

Preface

COSPAR PLANETARY PROTECTION POLICY FOR THE MARTIAN MOONS

An international consensus policy to prevent the biological cross-contamination of planetary bodies exists and is maintained by the Committee on Space Research (COSPAR) of the International Council for Science, which is consultative to the UN Committee on the Peaceful Uses of Outer Space. Currently, COSPAR's planetary protection policy does not specify the status of sample return missions from Phobos or Deimos, the moons of Mars. Although the moons themselves are not considered potential habitats for life or of intrinsic relevance to prebiotic chemical evolution, recent studies indicate that a significant amount of material recently ejected from Mars could be present on the surface of Phobos and, to a lesser extent, Deimos. Such interplanetary ejecta might mediate the transfer of viable organisms from one body to another. Such a process is sometimes referred to as lithopanspermia—a variant of the Arrhenius panspermia hypothesis.

Multiple space agencies, including the National Aeronautics and Space Administration (NASA), the European Space Agency (ESA), and the Japan Aerospace Exploration Agency (JAXA), are interested in plans for returning samples of material from the martian moons, which need to receive a planetary protection categorization of either restricted or unrestricted Earth return. A designation of restricted Earth return, per current NASA, ESA, JAXA, and COSPAR policy, would require samples to be maintained in high containment and undergo a biohazard test protocol after return. In addition, the moons of Mars are possible targets for future human exploration. Therefore, an understanding of the potential for life from Mars to persist on Phobos or Deimos is relevant to ensuring astronaut safety on those missions.

NASA and ESA rely on independent scientific advice from, respectively, the National Academies of Sciences, Engineering, and Medicine and the European Science Foundation (ESF) when faced with planetary protection questions not codified in current COSPAR policy. The National Academies and ESF have the ability to synthesize input from a wide spectrum of the scientific and technical communities and provide expert recommendations.

CREATION OF THE JOINT COMMITTEE

To lessen the scientific uncertainties concerning the planetary protection status of the martian moons, NASA and ESA commissioned researchers to perform modeling and experimental activities to assess the extent to which material from Mars might be deposited on the planet's moons and to assess the post-ejection environmental conditions that might inactivate potential martian life transported to Phobos and Deimos. The tests included hypervelocity impact sterilization of relevant Earth organisms, as well as the effects of ionizing radiation and heat on martian ejecta.

To provide an independent assessment of the results of these experimental activities, NASA and ESA issued parallel requests in 2016 to the National Academies' Space Studies Board (SSB) and ESF's European Space Science Committee (ESSC), respectively (see Appendix A). Both NASA and ESA specifically requested that the National Academies of Sciences, Engineering, and Medicine and the ESF jointly establish an ad hoc committee to review and assess recent research sponsored by NASA and ESA relating to the planetary protection concern that hypothetical martian life might exist on the surfaces of the martian moons, Phobos and Deimos, consequent to their ejection from the surface of Mars following a major impact event. Three specific tasks were enumerated (see the next section, Tasks 1, 2, and 6 in the statement of task).

Although there was no formal Japanese involvement in the commissioning of this study, it was generally agreed by NASA, ESA, the SSB, and ESSC that some participation by independent Japanese scientists was appropriate because of JAXA's plans to launch the Martian Moons Exploration (MMX) mission in the mid-2020s to collect and return samples from Phobos (or Deimos) to Earth. The joint National Academies-ESF Committee on the Planetary Protection Requirements for Sample Return Missions from the Martian Moons was formally established in mid-October 2017 and held its first and only planned meeting in London on November 6-9, 2017.

In March 2018, while the joint committee was assembling its draft report, NASA (with ESA concurrence) requested that the committee perform three actions (see Appendix B). First, delay the completion of its report. Second, plan to hold an additional meeting in fall 2018 to consider new results from ESA- and JAXA-sponsored groups studying the transfer of material from Mars to its moons. Third, expand the scope of its study by addressing three additional tasks (see Tasks 3 through 5 in the next section).

In fall 2018, five additional members were added to the committee to address the expanded scope of the study, and the committee met again in London on September 18-20, 2018.

The next section details the specific statements for the tasks (1-6) of the committee, in the context of their review of the ESA/NASA/JAXA research work.

STATEMENT OF TASK

The committee was specifically asked to address the following topics:

1. Review, in the context of current understanding of conditions relevant to inactivation of carbon-based life, recent theoretical, experimental, and modeling research on the environments and physical conditions encountered by Mars ejecta during the following processes:
   a. Excavation from the martian surface via crater-forming events;
   b. While in transit through cismartian space;
   c. During deposition on Phobos or Deimos; and
   d. After deposition on Phobos or Deimos.
2. Recommend whether missions returning samples from Phobos and/or Deimos should be classified as "restricted" or "unrestricted" Earth return in the framework of the planetary protection policy maintained by the ICSU Committee on Space Research (COSPAR);
3. In what specific ways is classification of sample return from Deimos a different case than sample return from Phobos?
4. What relevant information for classification of sample return is available from published studies of martian meteorites on Earth?
5. What are the planetary protection consequences of taking a surface sample at depths of 0–2 cm versus taking a sample extending down to depths of 2-10 cm or deeper?
6. Suggest any other refinements in planetary protection requirements that might be required to accommodate spacecraft missions to and sample returned from Phobos and/or Deimos.

REPORT REVIEW

A complete draft of the joint committee's report was assembled in October 2018 and sent to external reviewers on November 30. Responses to reviewer comments were drafted during the final week of December 2018 and a fully revised draft was approved for public release on January 10, 2019.

The work of the committee was made easier thanks to the important help, advice, and comments provided by numerous individuals from a variety of public and private organizations. These include the following: Allan Bennett (Public Health England), Catharine Conley (NASA), David Evans (Fluid Gravity Engineering Ltd.), Masaki Fujimoto (JAXA), Kazuhisa Fujita (JAXA), Yasuhiro Kawakatsu (JAXA), Gerhard Kminek (ESA), Kosuke Kurosawa (Chiba Institute of Technology), Manish Patel (The Open University), Victoria Pearson (The Open University), Mika Salminen (National Institute for Health and Welfare, Finland), J. Andrew Spry (SETI Institute), Thomas Statler (NASA), David Summers (Thales Alenia Space), Peter Triscott (Kallisto Consultancy), and Akihiko Yamagishi (Tokyo University of Pharmacy and Life Sciences). The committee offers special thanks to Kai Finster (Aarhus University) for his services as a consultant and participant in its first meeting.

The European Science Foundation elected not to conduct an independent review of this report. Rather, they agreed to abide by the report review policies and practices used by the National Academies. Therefore, this Consensus Study Report has been reviewed in draft form by individuals chosen for their diverse perspectives and technical expertise. The purpose of this independent review is to provide candid and critical comments that will assist the National Academies in making each published report as sound as possible and to ensure that the report meets the institutional standards for objectivity, evidence, and responsiveness to the study charge. The review comments and draft manuscript remain confidential to protect the integrity of the deliberative process.

We thank the following individuals for their review of this report: Kathrin Altwegg (University of Bern); Donna Blackmond, NAE (Scripps Research Institute); John Bridges (University of New Brunswick); Charles Cockell (University of Edinburgh); Gareth Collins (Imperial College); Dennis Discher, NAE, NAM (University of Pennsylvania); Katherine H. Freeman, NAS (Pennsylvania State University); Stephen Mackwell (American Institute of Physics); Ajay Malshe, NAE (University of Arkansas); John Spray (University of Leicester); and Erika Wagner (Blue Origin).

Although the reviewers listed above have provided many constructive comments and suggestions, they were not asked to endorse the conclusions or recommendations of this report nor did they see the final draft before its release. The review of this report was overseen by Steven J. Battel, NAE (Battel Engineering, Inc.). He was responsible for making certain that an independent examination of this report was carried out in accordance with the standards of the National Academies and that all review comments were carefully considered. Responsibility for the final content rests entirely with the authoring committee and the National Academies.

Errata:
John Bridges (University of Leicester);
John Spray (University of New Brunswick)

Contents

APPENDIXES

Summary

In 2016, the National Aeronautics and Space Administration (NASA) and the European Space Agency (ESA) issued parallel requests to the National Academies and the European Science Foundation (ESF), respectively, to undertake a study to determine the planetary protection classification of robotic sample return missions to the martian moons. In response to these requests to their parent organizations, the Space Studies Board (SSB) and the European Space Science Committee (ESSC) established a joint committee to address the requested tasks. (See "Statement of Task" in the Preface.)

Chapter 1 provides background to the statement of task and is organized in five sections: planetary protection policies, current understanding of the martian moons, background on martian meteorites, the Japan Aerospace Exploration Agency (JAXA) planned Martian Moons Exploration (MMX) mission, and a brief overview of research in support of MMX conducted by ESA (via the so-called SterLim team) and JAXA.

Chapter 2 contains a detailed overview of the work conducted in support of the planetary protection aspects of MMX by JAXA and the SterLim team sponsored by ESA. Chapter 2 also includes the committee's detailed critique and assessment of the research activities undertaken by the JAXA and SterLim teams.

Chapter 3 summarizes the committee's assessment of the JAXA and SterLim methodology, assumptions, and findings. This final chapter also investigates some additional arguments regarding planetary protection requirements for a sample return mission from the martian moons, and contains the committee's recommendations.

The first item in the committee's statement of task was as follows (see Preface):

1. Review, in the context of current understanding of conditions relevant to inactivation of carbon-based life, recent theoretical, experimental, and modeling research on the environments and physical conditions encountered by Mars ejecta during the following processes:
   a. Excavation from the martian surface via crater-forming events;
   b. While in transit through cismartian space;
   c. During deposition on Phobos or Deimos; and
   d. After deposition on Phobos or Deimos.

In this context, the committee reviewed the work of the SterLim and JAXA teams and issued the following findings:

- Even if life exists on Mars, the cell density and even its biochemical nature is unknown. Therefore, the value employed by the SterLim and JAXA teams is, as appropriate for a planetary protection calculation, a very conservative estimated based on current understanding of life as it exists in Mars-like extreme environments on Earth. (See "Potential Microbial Density on the Martian Surface" in Chapter 2.)
- The reason for the significant discrepancy in the amount of material transported to the martian moons as determined by the SterLim and JAXA teams could not be identified. Nevertheless, these uncertainties represent, in some sense, the current state of the art. (See "Mars Ejecta Formation and Transportation from the Martian Surface" in Chapter 2.)
- Shock heating during impacts is a highly localized process. When trying to resolve this adequately in numerical simulations, very high spatial resolutions are required. (See "Sterilization During Mars Ejecta Formation" in Chapter 2.)
- The survival rate during hypervelocity impacts cannot be determined based on the information available. However, the proposed survival rate of 10 percent is a reasonable estimate; albeit one lacking significant experimental evidence. (See "Sterilization During Mars Ejecta Formation" in Chapter 2.)
- The JAXA team's conclusion that particles smaller than 10 cm do not escape the martian atmosphere is not well supported. Therefore, subsequent analyses relying on this limit should be treated with care. (See "Sterilization by Aerodynamic Heating of Mars Ejecta" in Chapter 2.)
- The JAXA team's conclusion that aerodynamic heating of ejecta during passage through the martian atmosphere does not cause any significant sterilization is valid. (See "Sterilization by Aerodynamic Heating of Mars Ejecta" in Chapter 2.)
- The experimental hypervelocity impact data generated during the SterLim study is limited with respect to the large spectrum of possible impact conditions on the martian moons, could be biased, and is not conclusive. Given the small footprint of the data within the vast parameter space, extrapolations drawn from the experimental data currently available seemed ill-advised. SterLim's impact data was used to calibrate the exponential function used by the JAXA group to estimate and extrapolate the likely sterilization due to impact. (See Sterilization During Hypervelocity Impact on Phobos/Deimos Surfaces" in Chapter 2.)
- The estimations of the two teams as to the distribution and fate of Mars ejecta fragments deposited on the martian moons were based on different and limited experimental data. Therefore, a factor of uncertainty remains in the fraction deposited at the first impact. (See "Distribution of Mars Ejecta Fragments by Impacts, Recirculation, and Reimpact" in Chapter 2.)
- The SterLim team's use of aluminum, rather than a chemically inert surface, as a simulant environment for irradiation on Phobos/Deimos is problematic. In addition, the samples were irradiated in a frozen state, whereas the surface temperatures on the surfaces of the martian moons is frequently above the freezing point of water. (See "Sterilization by Radiation on Phobos/Deimos Surfaces" in Chapter 2.)
- Diurnal temperature cycling is an extremely significant factor in determining the survival of martian organisms deposited on the surfaces of Phobos or Deimos. Desiccation is bactericidal to even the most radiation-resistant microbes in a matter of months. (See "Sterilization by Radiation on Phobos/Deimos Surfaces" in Chapter 2.")
- The effect of meteoroid impacts following deposition of martian material on the surface of Phobos and Deimos has a minimal sterilizing effect due to the low flux of impactors. However, the fragmentation of ejecta due to the effects of thermal fatigue could significantly enhance the rate at which any organic matter present is degraded by exposure to the radiation. (See "Phobos/Deimos Surface Reformation by Natural Meteoroid Impacts" in Chapter 2.)

The second item in the committee's statement of task was as follows (see Preface):

2. Recommend whether missions returning samples from Phobos and/or Deimos should be classified as "restricted" or "unrestricted" Earth return in the framework of the planetary protection policy maintained by the ICSU Committee on Space Research (COSPAR).

A key factor in answering this question focused on whether or not an unidentified large (>10 km), young (<<1 million years) crater might exist on Mars. The committee finds that it is highly unlikely that such a large, young crater exists and has somehow escaped detection. (See Task 2 in Chapter 3.)

In determining whether samples returned from Phobos or Deimos should be classified as restricted or unrestricted Earth return, the committee considered the following factors:

- The work of the SterLim and JAXA teams can be considered as state of the art in regard to the modeling of the process of deposition of martian material on the surface of the martian moons. However, significant deficiencies exist in understanding, and there remain experimental and computational challenges associated with the quantitative estimation of ejecta mass and temperature distributions. Nevertheless, their work is convincing in showing that there is significant sterilization introduced throughout the chain of events. (See Task 2 in Chapter 3.)
- The issue of desiccation—as a result of diurnal thermal cycling on the surface of the martian moons—on any martian microbes was not considered by the SterLim and JAXA teams. At temperatures above the freezing point of water, desiccation is bactericidal to even the most radiation-resistant microbes in a matter of months. (See Task 2 in Chapter 3.)
- The relative influx of martian microbes from a Phobos/Deimos sample return mission versus the natural influx of direct Mars-to-Earth transfer can be shown to be smaller by several orders of magnitude. (See Task 2 in Chapter 3.)

Recommendation: After considering the body of work conducted by the SterLim and JAXA teams, the effect of desiccation on the surfaces of the martian moons, and the relative flux of meteorite- to spacecraft-mediated transfer to Earth, the committee recommends that samples returned from the martian moons be designated unrestricted Earth return.

The third item in the committee's statement of task was as follows (see Preface):

3. In what specific ways is classification of sample return from Deimos a different case than sample return from Phobos?

The different orbits and cross-sectional areas of Phobos and Deimos result in differences in the velocities associated with impacts of martian ejecta to their surfaces and in the total mass of material delivered to each moon. Both factors affect the total likelihood that microbes could survive delivery to the moons from Mars, and therefore raise the important question of whether Phobos and Deimos should be treated differently with respect to planetary protection requirements. While the studies conducted by the JAXA team did suggest that more martian material was likely to be present on Phobos than on Deimos, they also suggested that more organisms could theoretically survive transfer from Mars to Deimos. However, the latter conclusion was strongly dependent on the specific ejection geometries and velocities associated with modeling of a particular impact on Mars. (See Task 3 in Chapter 3.)

Recommendation: Given the uncertainty associated with impact sterilization assumptions, the committee recommends that Phobos and Deimos should not currently be treated differently in their planetary protection requirements.

The fourth item in the committee's statement of task was as follows (see Preface):

4. What relevant information for classification of sample return is available from published studies of martian meteorites on Earth?

An overview of the literature is included in Chapter 1 (see "Earth Inventory of Martian Meteorites"). The committee finds that the study of martian meteorites provides important context for studies of Mars and its moons and limited information (e.g., mass and flux to Earth) of relevance to planetary protection considerations. The unambiguous detection of an indigenous martian organism in a meteorite would be of great scientific and societal significance. (See Task 4 in Chapter 3.)

The fifth item in the committee's statement of task was as follows (see Preface):

> 5. What are the planetary protection consequences of taking a surface sample at depths of 0-2 cm versus taking a sample extending down to depths of 2-10 cm or deeper?

The committee identified two factors that could cause microbial survival probabilities to be different in these two depth ranges: ultraviolet irradiation and diurnal temperature cycling. Irradiation decreases microbe survival rates at the surface of Phobos or Deimos, but such radiation is attenuated within the top few millimeters of surface material. Therefore, this effect has no impact on sampling depth. Diurnal temperature changes are a significant factor in the top few cm. Therefore, samples from shallower depths on Phobos or Deimos have a lower risk for microbial contamination that those at a greater depth due to sterilization by thermal cycling. However, this additional factor is not needed to give confidence that samples from 2-10 cm depth will be below the established planetary protection limits for expected microbial contamination. (See Task 5 in Chapter 3.)

> **Recommendation: The committee recommends that no differences need to be made in planetary protection requirements for samples collected on the martian moons from depths 0-2 cm versus samples from 2-10 cm.**

The sixth item in the committee's statement of task was as follows (see Preface):

> 6. Suggest any other refinements in planetary protection requirements that might be required to accommodate spacecraft missions to and samples returned from Phobos or Deimos.

With respect to this last task, the committee limits its response to comments on three specific topics: uncertainty quantification, implications of the present work for Mars sample return missions, and the need to publish the work undertaken by the SterLim and JAXA teams.

Uncertainty quantification—The work of the SterLim and JAXA teams are prime examples of attempts to reach a specific conclusion about real-world activities based on combining the results from multiple numerical simulations and laboratory experiments. Each individual calculation or experiment is subject to various degrees of uncertainty. The science of uncertainty quantification seeks to determine the likelihood of specific outcomes for a system given that specific aspects of it are unknown or only weakly constrained. (See Task 6 in Chapter 3.)

> **Recommendation: The committee recommends that a significant effort be made by the planetary protection community to formally develop an uncertainty quantification protocol that can be used to estimate the cascading uncertainties that result from the integration of multiple computational models or other factors relevant to the quantitative aspects of planetary protection. Specific attention should be given to consideration of the significant uncertainties in the model inputs that exist because of limited available experimental or observational data.**

Implications for Mars sample return—What implications for a Mars sample return (MSR) mission can be drawn from this study and the work of the JAXA and SterLim teams? The main differences between MSR and Phobos/Deimos sample return missions are as follows:

- MSR sampling sites will be specifically selected to maximize sampling of evidence of extinct or extant life, whereas materials deposited on the martian moons originates from randomly distributed crater impact sites.

- Martian material present in a Phobos/Deimos sample would have undergone several physical sterilization processes (e.g., excavation by impact, collision with Phobos, and exposure to radiation), before it is actually sampled. Material collected on the surface of Mars will not have undergone such processes.
- MSR material might come from sites that mechanically cannot survive ejection from Mars and thus any putative life-forms would de facto not be able to survive impact ejection and transport to space. Such mechanical limitations do not apply for material collected on Mars.

Therefore, the committee finds that the content of this report and, specifically, the recommendations presented in it do not apply to future sample return missions from Mars itself. (See Task 6 in Chapter 3.)

Publication of the work of the SterLim and JAXA teams—The planetary protection, astrobiology, and planetary science communities would greatly benefit from the publication of the work undertaken by the SterLim and JAXA teams if for no other reason than to demonstrate the care and attention given to the investigation of planetary protection issues. (See Task 6 in Chapter 3.)

Recommendation: The committee recommends that the SterLim and JAXA teams formally publish the details of and results from their studies or make them readily available in some publicly accessible form.

1

Introduction

THE MARTIAN MOONS: PHOBOS AND DEIMOS

Mars, fourth planet from the Sun, is the outermost rocky planet of the inner solar system and marks the boundary between the terrestrial planets with solid surfaces and the giant planets beyond. Mars has two known moons: Phobos and Deimos. Phobos is roughly 22 km in diameter and orbits closer to Mars, with a semimajor axis of 9,377 km. Deimos is smaller—roughly 12 km in diameter—and orbits farther from Mars, with a semimajor axis of 23,460 km. The orbital periods of the two moons are very different, at 7.66 hours for Phobos and 30.35 hours for Deimos. Both moons are tidally locked, always presenting the same face toward Mars.

The orbits of both moons are unstable. Phobos's orbit lies inside the areosynchronous radius (i.e., the distance at which a martian satellite's orbital period is equal to one Mars day), and tidal forces are causing it to spiral in toward the planet on a time scale of 10-100 Ma. Deimos's orbit lies outside the areosynchronous radius, and tidal forces are causing it to spiral away from Mars on a similar time scale. As will become clear later (see "Sterilization by Radiation on Phobos/Deimos Surfaces," in Chapter 2), the time scale for orbital changes is significantly greater than that of relevance to the planetary protection issues being discussed in this report. Therefore, the effects of orbital changes can be ignored. A complete list of the physical characteristics of Phobos and Deimos is presented in Table 1.1.

No spacecraft mission has explored Phobos (Figure 1.1) or Deimos (Figure 1.2) as a primary objective, but several Mars-observing spacecraft have conducted remote, opportunistic observations of these bodies. In 1970, NASA's Mariner 7 first took pictures of Phobos silhouetted against Mars and revealed its small size, irregular shape, and dark surface. In 1971, NASA's Mariner 9 sent back the first images that were able to resolve surface features on Phobos and Deimos. Several other orbiting spacecraft have subsequently performed long-range observations, including the Viking 1 and 2 orbiters (NASA, 1970s and 1980s); the Phobos 2 mission (Soviet Union, 1988-1989); Mars Global Surveyor (NASA, 1997); the Mars Express mission (European Space Agency [ESA], since late 2003); NASA's Mars Reconnaissance Orbiter (NASA, since 2008); Mars Atmosphere and Volatile Evolution (MAVEN) mission (NASA, since 2014); Mars Orbiter Mission (Indian Space Research Organization, since 2014); and most recently, Mars Odyssey (NASA, Phobos observations since 2018). Russia attempted to send a sample return mission to Phobos—called Phobos-Grunt—in 2011, but the spacecraft failed to escape Earth's orbit and returned to Earth. While rovers and landers on the surface of Mars cannot get close to the moons, they have also provided some

TABLE 1.1 Phobos and Deimos Orbital Parameters

Property	Phobos	Deimos
Orbital semimajor axis	9,377 km	23,460 km
Orbital period	7.66 hr	30.35 hr
Orbital eccentricity	0.0151	0.0003
Orbital inclination, to Mars's equator	1.093 degrees	0.93 degree
Size	$26.06 \times 22.80 \times 18.28$ km^3	$15.0 \times 12.1 \times 10.4$ km^3
Density	1860 ± 13 kg m^{-3}	1490 ± 190 kg m^{-3}
Gravity	5.7×10^{-3} ms^{-2}	3×10^{-3} ms^{-2}
Normal reflectance, 0.55 μm	0.071 ± 0.012	0.068 ± 0.007
Estimated surface temperature range (min-max)	~150-300 K	~160-270 K

SOURCE: Table adapted from S.L. Murchie, P.C. Thomas, A.S. Rivkin, and N.L. Chabot, 2015, Phobos and Deimos, in *Asteroids IV* (P. Michel et al., eds.), University of Arizona, Tucson, doi:10.2458/azu_uapress_9780816532131-ch024.

Original source references: P.C. Thomas, 1993, Gravity, tides, and topography on small satellites and asteroids—Application to surface features of the martian satellites, *Icarus* 105:326; D.P. Simonelli, M. Wisz, A. Switala, D. Adinolfi, J. Veverka, P.C. Thomas, and P. Helfenstein, 1998, Photometric properties of Phobos surface materials from Viking images, *Icarus* 131:52-77; R.A. Jacobson, 2010, The orbits and masses of the martian satellites and the libration of Phobos, *Astronomical Journal* 139:668-679; K. Willner, X. Shi, and J. Oberst, 2014, Phobos's shape and topography models, *Planetary and Space Sciences* 102:52-59.

Temperature ranges references: J. Lunine, G. Neugebauer, and B. Jakosky, 1982, Infrared observations of Phobos and Deimos from Viking, *Journal of Geophysical Research* 87:B12; R.O. Kuzmin, T.V. Shingareva, and E.V. Zabalueva, 2003, An engineering model for the Phobos surface, *Solar System Research* 37:266; R.O. Kuzmin and E.V. Zabalueva, 2003, The temperature regime of the surface layer of the Phobos regolith in the region of the potential Fobos-Grunt Space Station landing site, *Solar System Research* 37:480; D.K. Lynch, R.W. Russell, R.J. Rudy, S. Mazuk, C.C. Venturini, H.B. Hammel, M.V. Sykes, R.C. Puetter, and R.B. Perry, 2007, Infrared spectra of Deimos (1-13 μm) and Phobos (3-13 μm), *Astronomical Journal* 134:4; J.L. Bandfield, S. Piqueux, T.D. Glotch, K.A. Shirley, T.C. Duxbury, J.R. Hill, C.S. Edwards, J.J. Plaut, V.E. Hamilton, and P.R. Christensen, 2018, "Mars Odyssey THEMIS Observations of Phobos: New Spectral and Thermophysical Measurements," Lunar and Planetary Science Conference, Abstract #2643.

disk-resolved images that show the moons' surfaces and have been useful in refining their ephemerides. A full summary of spacecraft exploration of Phobos and Deimos through 2014 is provided in a paper by Duxbury et al.[1]

The origins of Phobos and Deimos are unknown. Given their similarities in albedo, spectral properties, and density with carbonaceous material and D-type main belt asteroids, Phobos and Deimos were originally proposed to be captured objects from the inner or outer solar system.[2,3] However, the capture hypothesis is difficult to reconcile with the dynamics of Phobos's and Deimos's orbits. Formation from an accretion disk following a giant impact into early Mars or co-accretion in Mars orbit from Mars-like material can provide sufficient dissipation to damp the resulting debris disks down to the present orbital configuration and has been proposed as an alternative explanation for the moons' origins.[4,5,6,7,8] Estimates for the ages of the moons vary significantly depending on the formation model chosen.

---

[1] T.C. Duxbury, A. Zakharov, H. Hoffmann, and E.A. Guinness, 2014, Spacecraft exploration of Phobos and Deimos, *Planetary and Space Science* 102:9-17.

[2] K. Pang, J. Pollack, J. Veverka, A. Lane, and J. Ajello, 1978, The composition of Phobos: Evidence for carbonaceous chondrite surface from spectral analysis, *Science* 199:64-66.

[3] J.B. Pollack, J.A. Burns, and M.E. Tauber, 1979, Gas drag in primordial circumplanetary envelopes: A mechanism for satellite capture, *Icarus* 37:587-611.

[4] S.F. Singer, 1966, "On the Origin of the Martian Satellites Phobos and Deimos," pp. 317-321 in *Moon and Planets* (A. Dollfus, ed.), COSPAR Seventh International Space Science Symposium, COSPAR, Paris, France.

[5] R.A. Craddock, 2011, Are Phobos and Deimos the result of a giant impact?, *Icarus* 211:1150-1161.

[6] R.I. Citron, H. Genda, and S. Ida, 2015, Formation of Phobos and Deimos via a giant impact, *Icarus* 252:334-338.

[7] A.J. Hesselbrock and D.A. Minton, 2017, An ongoing satellite-ring cycle of Mars and the origins of Phobos and Deimos, *Nature Geoscience* 10:266-269.

[8] J.A. Burns, 1992, "Contradictory Clues as to the Origin of the Martian Moons," pp. 1283-1302 in *Mars* (H.H. Kieffer et al., eds.), University of Arizona, Tucson.

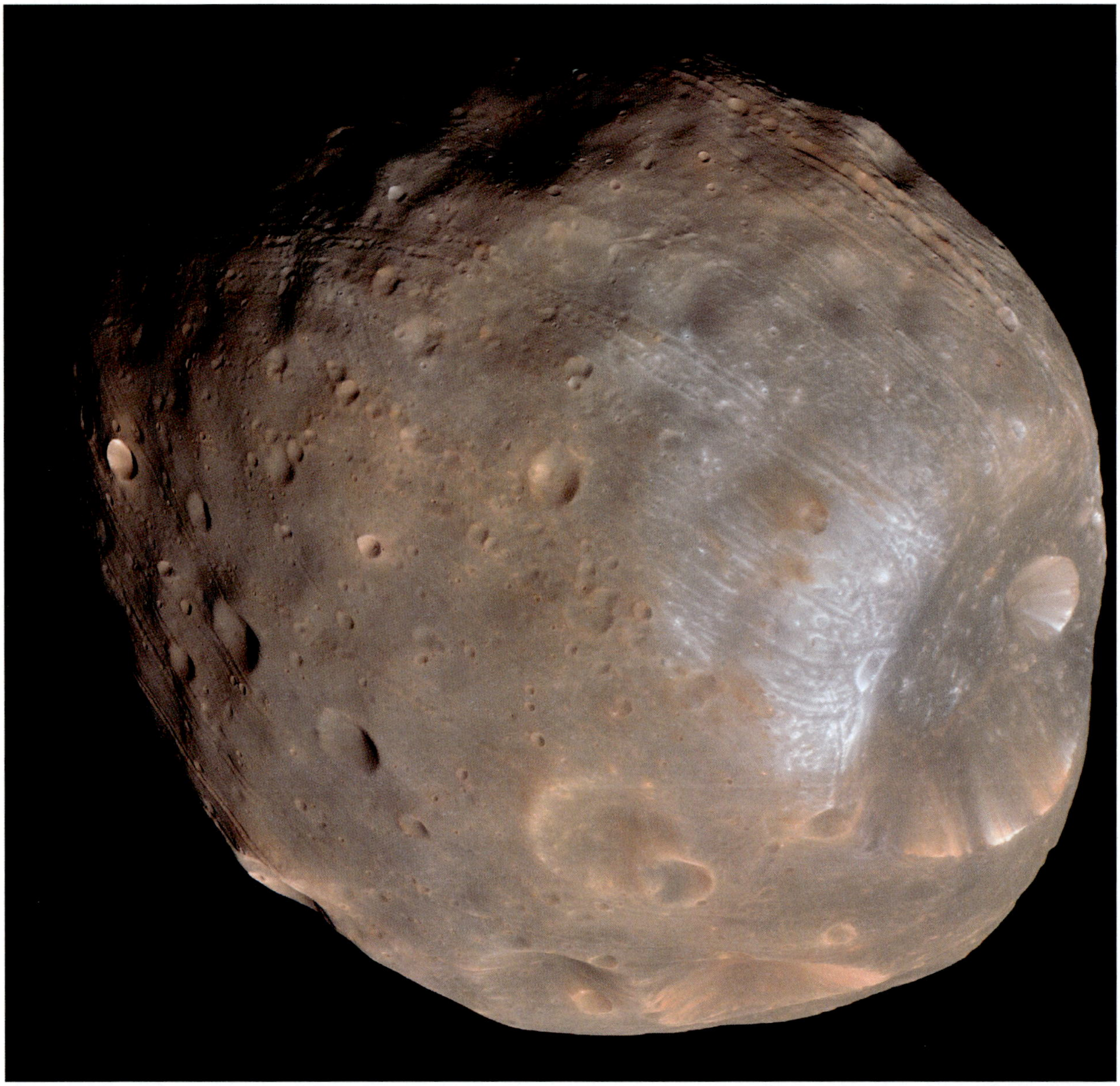

FIGURE 1.1 Phobos, the larger of Mars's two moons, as seen by the High-Resolution Imaging Sciences Experiment (HiRISE) on NASA's Mars Reconnaissance Orbiter in March 2008. The illuminated portion of the image is approximately 21 km across, and objects as small as approximately 6 m across can be resolved. SOURCE: Courtesy of NASA/JPL/University of Arizona.

Resolving the compositions of Phobos and Deimos will be a large step toward determining their origin. If the moons are captured bodies from the inner or outer solar system, they will probably resemble primitive meteorites or ordinary chondrites. If the moons formed via co-accretion or impact into differentiated Mars, they will probably resemble bulk Mars or differentiated basaltic martian crust.[9] Unfortunately, observations of Phobos's

[9] S.L. Murchie, P.C. Thomas, A.S. Rivkin, and N.L. Chabot, 2015, "Phobos and Deimos," in *Asteroids IV* (P. Michel et al., eds.), University of Arizona, Tucson, doi:10.2458/azu_uapress_9780816532131-ch024.

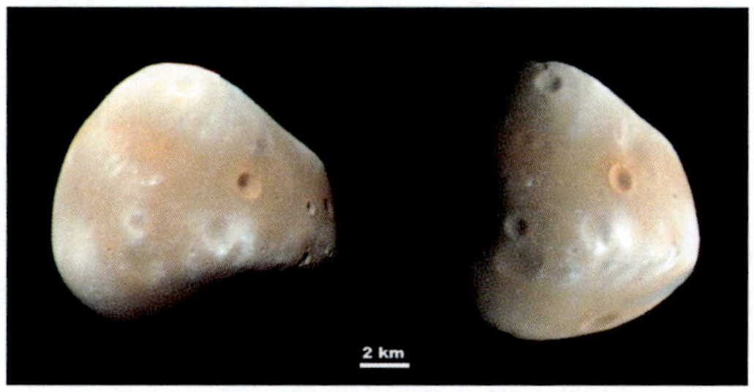

FIGURE 1.2 Deimos, the smaller of Mars's two moons, was imaged by the HiRISE camera on NASA's Mars Reconnaissance Orbiter on several occasions in February 2009. Features as small as approximately 60 m across can be resolved. SOURCE: Courtesy of NASA/JPL/University of Arizona.

and Deimos's composition to date have been ambiguous. Visible to near-infrared spectral data strongly suggest a chondritic composition,[10,11,12] while thermal infrared data suggest that a small basaltic component may be present.[13]

The spectral features suggesting a chondritic composition support the idea that the moons are captured asteroids. However, these features can be explained by exogenic processes such as the implantation of hydrogen from the solar wind. A strong argument against the capture hypothesis is based on the fact that each moon's orbit is almost circular and lies close to Mars's equatorial plane. Thus, the moons' current dynamical configurations would require the substantial dissipation of energy and angular momentum during the capture process. Explaining how this dissipation was achieved is difficult.

EARTH INVENTORY OF MARTIAN METEORITES

Important context for studies of Mars and its moons is provided by meteorites. Major impacts on Mars may deliver martian materials to Phobos and Deimos.[14,15] This process also delivers fragments of Mars—that is, martian meteorites—to Earth. The inventory of martian meteorites on Earth consists of about 115 volcanic and plutonic rocks whose chemical and oxygen–isotopic compositions differ from those of other meteorites and suggest their origin from differentiated parent bodies.[16] Sedimentary rocks that have been proven to exist on Mars—for example, from observations conducted by the Mars Global Surveyor spacecraft—are not among the martian meteorites identified so far. However, there may be additional complications in recognizing the martian origin of such samples, and also of other so far unknown materials from Mars, once they have been recovered on Earth.

Young igneous crystallization ages of 180 to 1,300 Ma for a large proportion of recognized martian meteorites point to derivation from a planet-size body, and martian atmosphere found trapped in impact-produced glass inclusions strongly advocate for their origin from Mars.[17,18] The majority of martian meteorites comprise shergottite

[10] S.L. Murchie and S. Erard, 1996, Spectral properties and heterogeneity of Phobos from measurements by *Phobos 2*, *Icarus* 123:63-86.

[11] A.S. Rivkin, R.H. Brown, D.E. Trilling, and J.F. Bell, 2002, Near-Infrared Spectrophotometry of Phobos and Deimos, *Icarus* 156(1):64-75.

[12] A.A. Fraeman, R.E. Arvidson, S.L. Murchie, A. Rivkin, T. Choo, J.-P. Bibring, B. Gondet, D. Humm, R.O. Kuzmin, N. Manaud, and E.V. Zabalueva, 2012, Analysis of disk-resolved OMEGA and CRISM spectral observations of Phobos and Deimos, *Journal of Geophysical Research* 117, doi:10.1029/2012JE004137.

[13] T.D. Glotch, C.S. Edwards, M. Yesiltas, K.A. Shirley, D.S. McDougall, A.M. Kling, J.L. Bandfield, and C.D.K. Herd, 2018, MGS-TES spectra suggest a basaltic component in the regolith of Phobos, *Journal of Geophysical Research: Planets* 123(10):2467-2484.

[14] L. Chappaz, H.J. Melosh, M. Vaquero, and K.C. Howell, 2012, "Transfer of Impact Ejecta Fragments Material from the Surface of Mars to Phobos and Deimos," Conference Paper AAS 12-212, 22nd AAS/AIAA Space Flight Mechanics Meeting, Univelt Inc., San Diego, Calif.

[15] K.R. Ramsley and J.W. Head III, 2013, Mars impact ejecta in the regolith of Phobos: Bulk concentration and distribution, *Planetary and Space Science* 87:115-129.

[16] H.Y. McSween, Jr., 1998, Martian meteorites, *Reviews in Mineralogy* 36:6.1-6.53.

[17] D.D. Bogard and D.H. Garrison, 1995, ^{39}Ar-^{40}Ar are of the Ibitira eucrites and constraints on the time of pyroxene equilibration, *Geochimica et Cosmochimica Acta* 59:4317-4322.

[18] A.H. Treiman, J.D. Gleason, and D.D. Bogard, 2000, The SNC meteorites are from Mars, *Planetary and Space Science* 48:1213-1230.

(>80 percent of all known martian meteorites), nakhlite (~10 percent), and chassignite (~3 percent) groups. Based on texture and mineralogy, shergottites are subdivided into basaltic-aphantic rocks with subequal proportions of plagioclase and pyroxene; olivine-phyric-aphanitic basalts with phenocrysts of olivine; and poikilitic-coarser grained rocks with oikocrysts of pyroxene enclosing olivine. Nakhlites are olivine clinopyroxenites, and chassignites are olivine-rich rocks called dunites. Additional Mars meteorite types include monomict orthopyroxenite breccia, Allan Hills (ALH) 84001,[19] and polymict regolith breccia Northwest Africa (NWA) 7034 and its pairs.[20,21,22] These breccias sample ancient martian crust, with an age of about 4.1 Ga for ALH 84001.[23] Zircons in NWA 7034 date to 4.428 Ga, with evidence of U-Pb disturbance at ~1.5-1.7 Ga.[24] Two recently recognized meteorites, NWA 8159 and NWA 7635, expand shergottite types, sampling a discrete igneous unit from the early Amazonian (~2.3-2.4 Ga).[25,26] Despite identification of sedimentary rocks on Mars through satellite- and rover-derived morphological observations, there are no such rocks represented within current meteorite collections.

All martian meteorites on Earth were ejected from Mars by hypervelocity impact, originating within a near-surface "spall" zone of inverted pressure gradient, caused by interference between shock waves and rarefactions near the free surface.[27] This spall zone comprises accelerated solid rock and has been studied both numerically and analytically.[28,29,30,31] Ejection ages indicate that the martian meteorites were delivered to Earth by less than eight discrete impact events between 0.7 and 20 Ma ago.[32] Attempts have been made to identify meteorite source craters using spectral matching.[33,34] However, such efforts have been hampered by dust that obscures primarily the youngest igneous terrains such as Tharsis.[35] The bias of martian meteorites toward young igneous rocks has been investigated through computer simulation by Head et al.[36] Their results show that the size of the ejected fragments is affected by target strength; weaker materials, like sedimentary rocks, require larger, and therefore rarer, impact events. This observation may account for the paucity of breccias in the current collection and the absence of sedimentary martian meteorites. There may be added complications in recognizing the martian origin of these samples once they have been recovered on Earth. Hypervelocity impact into coherent targets, such as

---

19 D.W. Mittlefehldt, 1994, ALH 84001: A cumulate orthopyroxenite member of the martian meteorite clan, *Meteoritics* 29:214-221.

20 C. Agee, N.V. Wilson, F.M. McCubbin, K. Ziegler, V.J. Polyak, Z.D. Sharp, Y. Asmerom, M.H. Nunn, R. Shaheen, M.H. Thiemens, A. Steele, et al., 2013, Unique meteorite from early Amazonian Mars: Water-rich basaltic breccia Northwest Africa 7034, *Science* 339:780-785.

21 M. Humayun, A. Nemchin, B. Zanda, R.H. Hewins, M. Grange, A. Kennedy, J.P. Lorand, C. Gopel, C. Fieni, S. Pont, and D. Deldicque, 2013, Origin and age of the earliest martian crust from meteorite NWA 7533, *Nature* 503:513-516.

22 A. Wittman, R.L. Korotev, B.L. Joliff, A.J. Irving, D.E. Moser, I. Barker, and D. Rumble III, 2015, Petrography and composition of martian regolith breccia meteorite Northwest Africa 7475, *Meteoritics and Planetary Science* 50:326-352.

23 J.J. Bellucci, A.A. Nemchin, M.J. Whitehouse, J.F. Snape, P. Bland, and G.K. Benedix, 2015, The Pb isotopic evolution of the martian mantle constrained by intial Pb loss in martian meteorites, *Journal of Geophysical Research: Planets* 120:2224-2240.

24 M. Humayun, A. Nemchin, B. Zanda, R.H. Hewins, M. Grange, A. Kennedy, J.P. Lorand, C. Gopel, C. Fieni, S. Pont, and D. Deldicque, 2013, Origin and age of the earliest martian crust from meteorite NWA 7533, *Nature* 503:513-516.

25 C.D.K. Herd, E.L. Walton, C.B. Agee, N. Muttik, K. Zeigler, C.K. Shearer, A.S. Bell, et al., 2017, The Northwest Africa 8159 martian meteorite: Expanding the martian sample suite to the early Amazonian, *Geochimica et Cosmochimica Acta* 218:1-26.

26 T.J. Lapen, M. Righter, R. Andreasen, A.J. Irving, A.M. Stakoski, B.L. Beard, K. Nishiizumi, A.J.T. Jull, and M.W. Caffee, 2017, Two billion years of magmatism recorded from a single Mars meteorite ejection site, *Science Advances* 3:6.

27 H.J. Melosh, 1985, Ejection of rock fragments from planetary bodies, *Geology* 13:144-148.

28 P.H. Warren, 1994, Lunar and martian meteorite delivery services, *Icarus* 111:338-363.

29 H.J. Melosh, 1995, Cratering dynamics and the delivery of meteorites to the Earth, *Meteoritics* 30:545-546.

30 J.N. Head, H.J. Melosh, and B.A. Ivanov, 2002, Martian meteorite launch: High-speed ejecta from small craters, *Science* 298:1753-1756.

31 N. Artemieva and B. Ivanov, 2004, Launch of martian meteorites in oblique impacts, *Icarus* 171:84-101.

32 L.E. Nyquist, D.D. Borg, C.-Y. Shih, D. Greshake, D. Stöffler, and O. Eugster, 2001, Ages and geologic histories of martian meteorites, *Space Science Review* 96:105-164.

33 A. Ody, F. Poulet, C. Quantin, J.P. Bibring, J.L. Bishop, and M.D. Dyar, 2015, Candidate source regions of martian meteorites as identified by OMEGA/MEx, *Icarus* 258:366-383.

34 V.E. Hamilton and P.R. Christensen, 2003, High spectral and spatial resolution analyses of martian meteorite-like compositions on the surface of Mars, *Meteoritics and Planetary Science* 38:76.

35 N.P. Lang, L.L. Tornabene, H.Y. McSween, Jr., and P.R. Christensen, 2009, Tharsis-sourced relatively dust-free lavas and their possible relationship to martian meteorites, *Journal of Volcanology and Geothermal Research* 185:103-115.

36 J.N. Head, H.J. Melosh, and B.A. Ivanov, 2002, Martian meteorite launch: High-speed ejecta from small craters, *Science* 298:1753-1756.

shergottite-nakhlite-chassignite source terrains, may eject decimeter-size rocks, leaving their trace as craters as small as 3 km diameter.[37]

This same process of impact spallation may also eject martian materials into Phobos- and Deimos-crossing trajectories,[38,39] necessitating further assessment of the amount of martian material on these moons, under the auspices of planetary protection (this study). Ejecta arriving directly to Phobos from Mars (referred to as primary ejecta), intersects the surface at ~2-3 km/s. Due to the moon's small size, and therefore low escape velocity (~4-10 m/s), a large amount of secondary ejecta (~95-99 percent) that is temporarily inserted into martian orbit may subsequently reaccrete on the moon. The reaccretion interval for secondary ejecta to Phobos ranges from several days to hundreds of years.[40] Based on these models, the amount of martian material in the regolith of Phobos was computed to be ~75 ppm in the last 10 Ma and ~250 ppm delivered during the last 3.5 Ga. This material is primarily within 0.4-1.0 m of the surface, with 10 to 60 times less in terms of bulk concentration in deeper, and therefore older (>500 Ma), regolith units. The process of delivery of Mars's material to its moons has been revisited by the Japan Aerospace Exploration Agency (JAXA), and a synopsis of their results can be found in Chapter 2.

During impact ejection, the rock fragments—some destined to become meteorites— are shock-metamorphosed. Shock effects in martian meteorites are recorded as petrographically observable features in constituent minerals, including mechanical deformation and transformation. Transformation of plagioclase to a diaplectic glass called maskelynite is sensitive to composition (Ca-content) and shock pressure, which has been calibrated by shock-recovery experiments.[41] All martian meteorites record shock effects, and their study can be used to estimate shock pressure and post-shock temperature.[42] The shock-induced temperature increase is governed by the pressure-volume work achieved by the shock wave, which may be estimated using the linear relation of shock wave and particle velocity across specific pressure intervals, as described in a 2005 paper by Fritz et al.[43] Studies of shock effects in martian meteorites show that they have experienced a range of shock conditions, from weakly shocked nakhlites (5-10 GPa) to more strongly shocked shergottites (20-55 GPa). These pressure estimates are based on mineral deformation in olivine and pyroxene, including, but not limited to, planar fractures, undulose extinction, planar deformation features and mechanical twinning (pyroxene only), and complete to partial transformation of plagioclase to maskelynite. Calculated post-shock temperature increase (ΔT) ranges from 10 ± 20 K in nakhlites, to 50 ± 5 K at the lower end of shock in shergottites (20 GPa; Yamato-980459), to 800 ± 200 K at the upper limit (55 GPa; ALH 77005).[44] Low post-shock temperatures are supported by study of ALH 84001 magnetization, demonstrating that this meteorite did not realize temperatures greater than 313.15 K since its formation.[45] These shock conditions apply to those experienced by the bulk rock; however, during shock, shear zones may develop and open spaces (e.g., cracks, fractures, vesicles) collapse, forming hot spots within the rock and generating small volumes of shock-produced melt. The localized temperature conditions within shock melt may be in excess of 1500-2000 K; however, these represent small volumes of melt (<1 to ~3 percent) that are heterogeneously distributed throughout the sample.[46]

---

37 Ibid.

38 L. Chappaz, H.J. Melosh, M. Vaquero, and K.C. Howell, 2012, "Transfer of Impact Ejecta Fragments Material from the Surface of Mars to Phobos and Deimos," 22nd AAS/AIAA Space Flight Mechanics Meeting. AAS Conference paper 12-212.

39 K.R. Ramsley and J.W. Head III, 2013, Mars impact ejecta in the regolith of Phobos: Bulk concentration and distribution, *Planetary and Space Science* 87:115-129.

40 Ibid.

41 D. Stöffler, C. Meyer, J. Fritz, G. Horneck, R. Möller, C. Cockell, S. Ott, J.P. de Vera, U. Hornemann, and N.A. Artemieva, 2006, Impact Experiments in Support of "Lithopanspermia": The route from Mars to Earth, Abstract 1551, 37th Lunar and Planetary Science Conference, https://www.lpi.usra.edu/meetings/lpsc2006/pdf/1551.pdf.

42 See, for example, J. Fritz, N. Artemieva, and A. Greshake, 2005, Ejection of martian meteorites, *Meteoritic and Planetary Science* 40:1393-1411, and references therein.

43 J. Fritz, N. Artemieva, and A. Greshake, 2005, Ejection of martian meteorites, *Meteoritic and Planetary Science* 40:1393-1411.

44 Ibid.

45 B.P. Weiss, J.L. Kirschvink, F.J. Baudenbacher, H. Vali, N.T. Peters, F.A. MacDonald, and J.P. Wikswo, 2000, A lower temperature transfer of ALH 84001 from Mars to Earth, *Science* 290:791-795.

46 E.L. Walton and C.S.J. Shaw, 2009, Understanding the textures and origin of shock melt pockets in martian meteorite from petrographic studies, comparisons with terrestrial mantle xenoliths, and experimental studies, *Meteoritics and Planetary Science* 44:55-76.

There is considerable debate about the presence, origin, and meaning of organic material in martian meteorites. Chains of tiny magnetite crystals associated with carbonate globules found in ALH 84001 have been interpreted as evidence for possible ancient biological activity on Mars.[47] However, the magnetite may also have formed by inorganic processes—for example, thermal decomposition of carbonates during shock heating.[48] Also, organic compounds reported from ALH 84001 and EETA 79001 appear to be of terrestrial and not martian origin.[49] More recently, kerogen-like organic matter present in the 2011 fall, Tissint, as well as methane released from six martian meteorites, have been taken as hints of biological activity.[50,51] Nakhlites, the least shock-metamorphosed martian meteorites, contain various alteration assemblages—for example, clay minerals, sulfates, and halite—attesting to interaction between martian crustal fluids and the parent igneous rocks.[52] In strongly shocked shergottites, geochemical signatures of martian alteration such as D- and Cl-enrichment are found preferentially in quenched shock melt.[53] Despite these detailed studies in search of evidence for biological activity, there has been no unambiguous evidence for early life found in martian meteorites. This, however, does not generally preclude the possibility of sampling martian material that may contain signs of biological activity.

Chapter 3 returns to the discussion of martian meteorites, because they will prove to be an important factor in determining whether or not samples from the martian moons are designated restricted or unrestricted Earth return.

FUTURE MISSIONS TO THE MARTIAN MOONS: THE MMX MISSION

The preceding two sections have demonstrated that Phobos and Deimos are high-priority targets for a future dedicated spacecraft mission, especially one that could return samples for detailed study in terrestrial laboratories.[54] Resolving the questions of the moons' origins will advance researchers' understanding of how planetary systems form. Studying material from Phobos and Deimos will also provide information about primordial material transport in the earliest period of solar system history. If the moons are captured bodies originating from the outer solar system, they would provide important clues about material transport across the snow line marking the frontier between the inner and outer solar system.

The Martian Moons Exploration (MMX) mission is a robotic spacecraft mission under development by JAXA for launch in September 2024. MMX will be a 5-year sample return mission with the following mission profile:

- September 2024—Launch
- August 2025—Arrive at Mars
- 2026—Observation of Phobos for landing site selection
- 2026 or 2027—Proximity phase: landing on Phobos for sampling
- August 2028—Depart from Mars
- July 2029—Arrive at Earth

---

[47] D.S. McKay, J.E.K. Gibson, K.L. Thomas-Keprta, H. Vali, C.S. Romanek, S.J. Clemett, X.D.F. Chiller, C.R. Maechling, and R.N. Zare, 1996, Search for past life on Mars: Possible relic biogenic activity in martian meteorite ALH 84001, *Science* 273:924-930.

[48] A.J. Brearley, 2003, Magnetite in ALH 84001: An origin by shock-induced thermal decomposition of iron carbonate, *Meteoritics and Planetary Science* 38:849-870.

[49] A.J.T. Jull, C. Courtney, D.A. Jeffrey, and J.W. Beck, 1998, Isotopic evidence for a terrestrial source of organic compounds found in martian meteorites Allan Hills 84001 and Elephant Moraine 79001, *Science* 279:366-369.

[50] Y. Lin, A. El Goresy, S. Hu, J. Zhang, P. Gillet, Y. Xu, J. Hao, et al., 2014, NanoSIMS analysis of organic carbon from the Tissint martian meteorite: Evidence for the past existence of subsurface organic-bearing fluids on Mars, *Meteoritics and Planetary Science* 49:2201-2218.

[51] N.J.F. Blamey, J. Parnell, S. McMahon, D.F. Mark, T. Tomkinson, M. Lee, J. Shivak, M.R.M. Izawa, N.R. Banerjee, and R.L. Flemming, 2015, Evidence for methane in martian meteorites, *Nature Communication*, doi:10.1038/ncomms8399.

[52] J.C. Bridges, D.C. Catling, J.M. Saxton, T.D. Swindle, I.C. Lyon, and M.M. Grady, 2001, Alteration assemblages in martian meteorites: Implications for surface-near processes, *Space Science Reviews* 96:365-392.

[53] C.R. Kuchka, C.D.K. Herd, E.L. Walton, Y. Guan, and Y. Liu, 2017, Martian low-temperature alteration materials in shock-melt pockets in Tissint: Constraints on their preservation in shergottite meteorites, *Geochimica et Cosmochimica Acta* 210:228-246.

[54] See, for example, S.L. Murchie, D.T. Britt, and C.M. Pieters, 2014, The value of Phobos sample return, *Planetary and Space Sciences* 102:176-182.

TABLE 1.2 Instruments to Be Carried by the MMX Mission

Instruments	Objectives	Specifications
TENGOO (Telescopic Nadir Imager for Geomorphology)	Geological features	Field of view (FOV): 1.1 × 0.82 degrees Spatial resolution: ~40 cm at 20 km altitude
OROCHI (Optical Radiometer Composed of Chromatic Imagers)	Geological features Hydrated minerals Space weathering	FOV: 66 × 53 degrees Wavelength: 390, 480, 550, 650, 700, 860, 950 nm Spatial resolution: 20 m at 20 km altitude
MacrOmega (Macroscopique Observatoire pour la Minéralogie, l'Eau, le Glaces et l'Activité)	Hydrated minerals Water molecules Organic materials	FOV: 6 degrees Wavelength: 0.9-3.6 μm Spatial resolution: <20 m at 20 km altitude
Lidar (Light Detection and Ranging)	Topographic features	Ranging distance: 100 m-100 km Ranging resolution: 0.5 m
MEGANE (Mars Moon Exploration with Gamma Rays and Neutrons)	Major element composition	Gamma-ray energy: 0.4-8 MeV Energy resolution: <5.1 keV (FWHM) at 1454 keV Neutron energy: thermal, epithermal, and fast (0.01-7 MeV)
MSA (Mass Spectrum Analyzer)	Space ion environment Possible ice inside Phobos	Ion energy: 10-30 keV/q Energy resolution: $\Delta E/E \sim 20$ percent Ion mass: 1-60 amu Mass resolution: $M/\Delta M \sim 100$

NOTE: TENGOO, MacrOmega (Macroscopic Observatory for Mineralogy, Water, Ice and Activity; contributed by the French Space Agency, CNES), and MEGANE (contributed by NASA) will play extremely important roles in the landing site selection on Phobos. In particular, TENGOO's high-resolution imaging ability will be crucial to landing safely. FWHM, full width at half maximum.

MMX has three scientific objectives.[55] They are, in priority order:

1. To understand the origin of the martian moons. Are Phobos and Deimos captured primordial asteroids or leftover accreted debris from a significant impact in Mars's history?
2. To make progress in understanding planetary system formation and primordial material transport of material between the inner and outer portions of the early solar system.
3. To understand processes in cismartian space and to investigate how it might have changed in response to the evolution of the surface and atmosphere of Mars throughout the history of the solar system.

The science goals of MMX are to be addressed with a comprehensive suite of instruments (Table 1.2) and two sampling systems, JAXA's C-Sampler (a coring device) and NASA's P-Sampler (a pneumatic device).

P-Sampler

The P-Sampler, one of NASA's major contributions to the MMX mission, is a pneumatic sample collection device mounted on one of the footpads of the spacecraft's landing legs (Figure 1.3). The main characteristics of its operation are as follows:

- Samples only the top 1 cm of Phobos's regolith.
- Uses gas pressure to agitate surface material and blow it to the Sample Canister.

[55] H. Miyamoto, "Japanese Mission of the Two Moons of Mars with Sample Return from Phobos," presentation to Mars Program Assessment Group, March 17, 2016, https://mepag.jpl.nasa.gov/meetings/2016-03/17_Miyamoto.pdf.

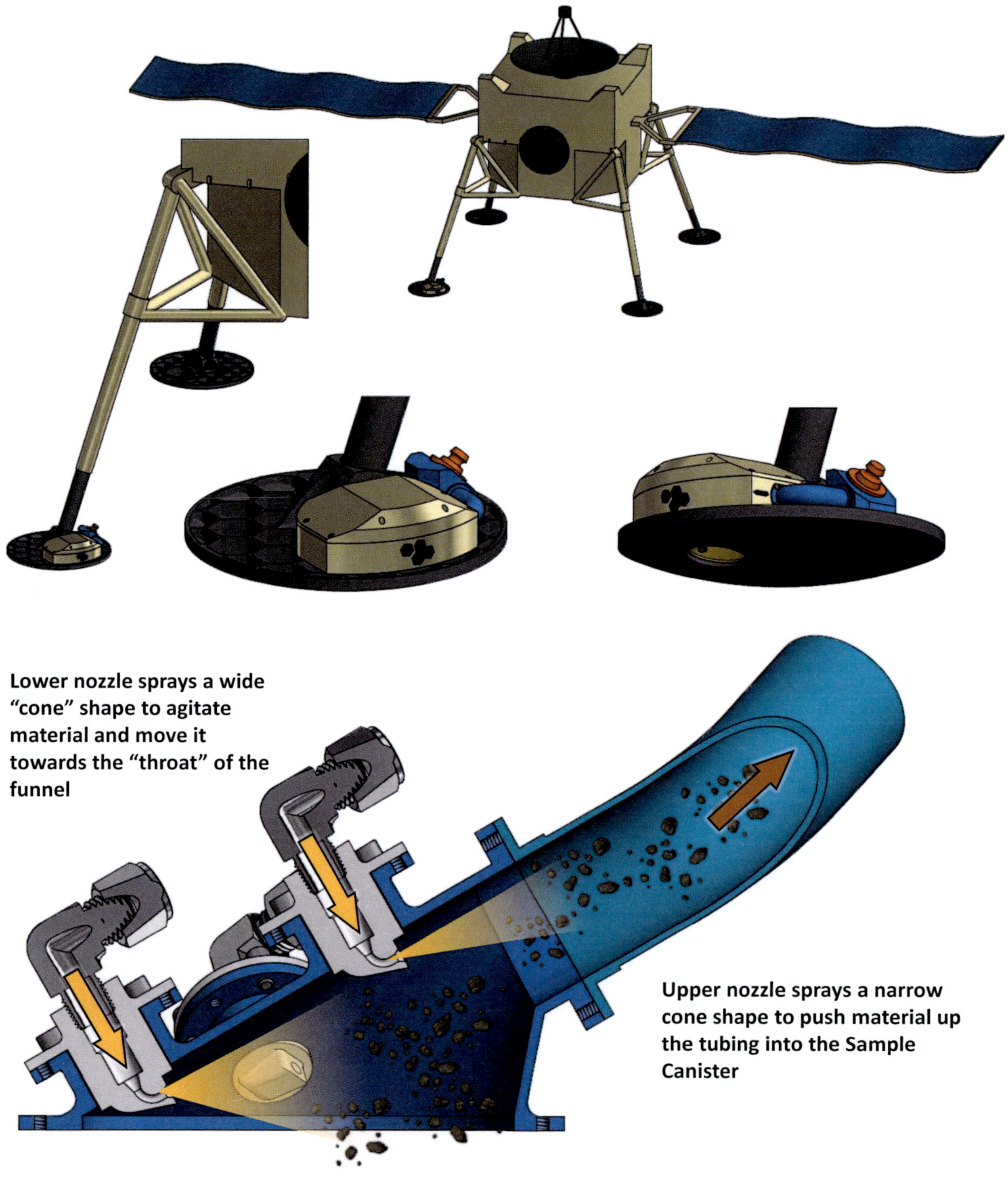

FIGURE 1.3 *Top:* The location of P-Sampler (beige box) and its Sample Canister (blue box) on one of MMX's footpads. The dark circle on the near side of the spacecraft is the Sample Return Capsule. *Bottom:* View of the nozzles through which high-pressure gas emerges. The gas agitates the top 1 cm of regolith and sprays it into the removable Sample Canister. Once a sample has been collected, a robotic arm (not shown) located on the underside of the spacecraft detaches the Sample Canister from the P-Sampler and transfers it to the Sample Return Capsule. SOURCE: Images courtesy of Honeybee Robotics Spacecraft Mechanisms Corporation.

- Once a sample has been collected, a robotic arm, mounted on the underside of the spacecraft moves the Sample Canister to the Sample Return Capsule located on the side of the spacecraft.

C-Sampler

The C-Sampler is mounted on the underside of the MMX spacecraft (Figure 1.4); it consists of three separate coring bits designed to retrieve samples from two different locations. The third bit is a spare. The general characteristics of the C-Sampler are as follows:

- Each core tube has an inside diameter of 2.5 cm and is 6 cm long.
- 10 g of material is gathered for each corer at a depth of greater than 2 cm.
- The corer is located on the end of a robotic arm attached to the underside of the spacecraft.
- Three-dimensional imaging is used to determine the best location for sampling.
- Once a core sample is collected, the robotic arm transfers the sample tube to the Earth-return capsule mounted on the side of the spacecraft.

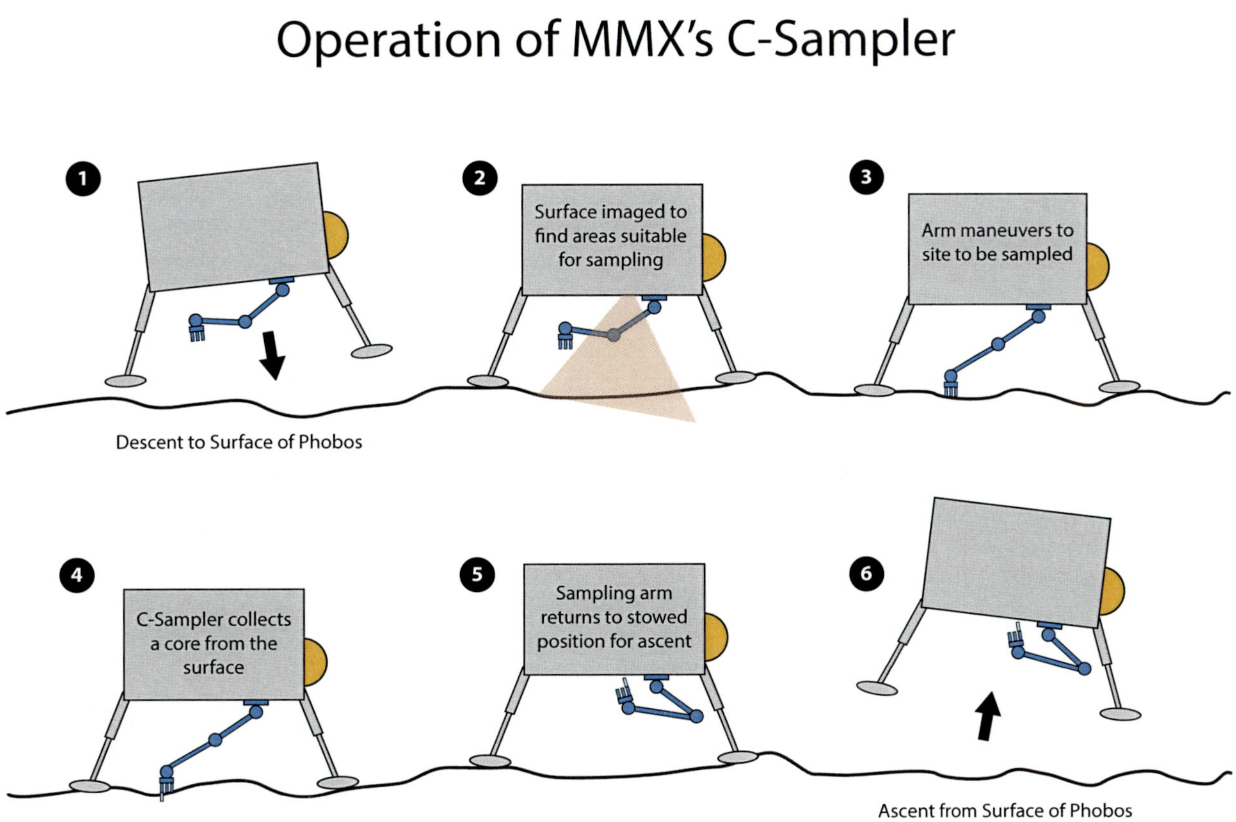

FIGURE 1.4 The C-Sampler uses a coring device on the end of a robotic arm to acquire a sample. Once collected, the arm transfers the sample in the core tube to the Sample Return Capsule (shown in orange) mounted on the right-hand side of the spacecraft.

Sample Return Capsule

The Sample Return Capsule is located on the side of the MMX spacecraft. Its design is based on that used for JAXA's Hayabusa 1 and 2 asteroid sample return missions. Following the completion of sample acquisition and remote-sensing studies of the martian moons, the MMX spacecraft, with Sample Return Capsule still attached, uses its ion propulsion system to return to Earth. Just prior to entry into Earth's atmosphere, the Sample Return Capsule detaches from MMX. The MMX spacecraft burns up in Earth's atmosphere. But, the Sample Return Capsule, protected by its heatshield and slowed by a parachute, drifts down to a soft landing in the Woomera Prohibited Area, in South Australia.[56] The specific characteristics of the Sample Return Capsule are as follows:

- Capsule diameter—60 cm (Hayabura 2, 40 cm)
- Capsule mass—38 kg (Hayabusa 2, 16.5 kg)
- Separation mechanism mass—7 kg
- Total subsystem mass—45 kg
- Payload volume—15 × 15 × 15 cm^3
- Total sample mass—0.01-0.03 kg
- Thermal protection system—carbon phenolic

PLANETARY PROTECTION AND COSPAR POLICY

Planetary protection policies have a twofold goal, as follows:[57]

- The control of forward contamination in the form of viable microbial life from Earth; and
- The control of back contamination by extraterrestrial materials collected and returned to the Earth-Moon system by spacecraft missions.

The rationales for these goals are also twofold:[58]

- To preserve the integrity of Earth's biosphere; and
- To protect the biological and environmental integrity of other solar system bodies for future science missions, especially those relating to the origins of life and prebiotic chemical evolution.

The 1967 United Nations (UN) Outer Space Treaty (OST),[59] to which most spacefaring nations are signatory, includes the following language as part of Article IX: "States Parties to the Treaty shall pursue studies of outer space, including the Moon and other celestial bodies, and conduct exploration of them so as to avoid their harmful contamination, and also adverse changes in the environment of the Earth resulting from the introduction of extraterrestrial matter and, where necessary, shall adopt appropriate measures for this purpose." In addition, Article VI of the same treaty specifies the following: "States Parties to the Treaty shall bear international responsibility for national activities in outer space, including the Moon and other celestial bodies, whether such activities are carried on by governmental agencies or by non-governmental entities."

Technical aspects of planetary protection policies are developed by individual space agencies and coordinated through the Committee on Space Research (COSPAR), part of the International Council of Science (ICSU). International planetary protection consensus guidelines are developed through a harmonization process conducted by

[56] For additional information about the Woomera Prohibited Area, see, for example, http://www.defence.gov.au/woomera/about.htm.

[57] See, for example, National Academies of Sciences, Engineering and Medicine, *Review and Assessment of Planetary Protection Policy Development Processes,* The National Academies Press, Washington, DC, 2018, p. 9.

[58] Ibid.

[59] Treaty on Principles Governing the Activities of States in the Exploration and Use of Outer Space, Including the Moon and Other Celestial Bodies, opened for signature January 27, 1967, 18 U.S.T. 2410, 610 U.N.T.S. 25.

COSPAR's Panel on Planetary Protection (PPP). The UN Committee on Peaceful Uses of Outer Space has accepted COSPAR's planetary protection policy as a guide for compliance with the OST.

COSPAR's deliberations occur regularly, with participants reporting new scientific findings with policy implications (e.g., water being more abundant at a particular target than was previously recognized) and raising questions regarding specific concerns (e.g., new activities in space exploration that could affect policy compliance). The PPP develops recommendations that the COSPAR Bureau and Council may adopt for inclusion into the official COSPAR planetary protection policy.[60] Through this process, the COSPAR planetary protection policy has evolved steadily and incrementally over the years since it was initially created. Space agencies such as NASA, the European Space Agency (ESA), and JAXA formulate and implement planetary protection policies and procedures that are consistent with the COSPAR planetary protection policy.

NASA and ESA maintain their respective planetary protection policies, administer associated procedures to ensure compliance with them, and oversee compliance with formal implementation requirements that are assigned to each mission. Agency policies are informed by the most current scientific information available about the target bodies and about life on Earth.

Planetary protection policies are not static but evolve over time based on the increasing knowledge and understanding of both planetary environments and the physical and chemical limits of terrestrial life. Conclusions and recommendations generated by internal and external advisory groups chartered by space agencies such as NASA and ESA are weighted and assessed in an iterative manner by COSPAR's PPP. Consensus policy recommendations developed by the PPP are then forwarded for discussion and ultimate approval by COSPAR's Bureau and Council prior to becoming official COSPAR policy. The development of the concept of Special Regions on Mars is a good example of how planetary protection policies are developed and evolve as new information becomes available.[61]

COSPAR planetary protection policy sets requirements for each spacecraft mission and target body depending on the type of encounter it will have (e.g., flyby, orbiter, or lander) and the nature of its destination (e.g., a planet, a moon, a comet, or an asteroid). If the target body has the potential to provide clues about the origins and evolution of life or prebiotic chemical evolution, spacecraft going there are required to meet a higher level of cleanliness, and some operating restrictions will be imposed. Spacecraft going to target bodies with the potential to support Earth life undergo stringent cleaning and bioload-reduction processes, up to and including subjecting the entire spacecraft to a dry heat microbial reduction (heating to ~385 K for 30 hours) or equivalent process. Such missions may also be subject to operating restrictions.

The fundamental challenge for those drafting planetary protection policies and their implementations is to craft requirements that are conservative. In practice, conservatism means that when assessing the risks posed by forward or back contamination, unknown or poorly known factors are overestimated if potentially harmful or underestimated if potentially beneficial. However, requirements should not be so conservative as to preclude the design or operation of a spacecraft mission designed to explore a planetary body of scientific interest and planetary protection concern.

The MMX mission presents an interesting test case of balancing conservatism against practicality. Phobos and Deimos are not in themselves objects of interest to studies of the origins of life and prebiotic chemical evolution. Therefore, spacecraft missions to these bodies present little to no chance that any onboard biological contamination from Earth will compromise future scientific investigations. Such missions would be subject to only the most minimal of requirements (e.g., documentation as to where the spacecraft went and what it did) relating to the control of forward contamination.

Returning samples from Phobos and Deimos is a more difficult question. Extraterrestrial samples returned to Earth by spacecraft missions are subject to more- or less-rigorous inflight and post-return containment restrictions depending on whether the body from which they are collected is designated "restricted" or "unrestricted" Earth return, respectively. The latter designation is reserved for spacecraft missions returning materials from

---

[60] G. Kminek, C. Conley, V. Hipkin, and H. Yano, 2017, COSPAR Planetary Protection Policy, *Space Research Today* 200:12-25.

[61] See, for example, National Academies of Sciences, Engineering and Medicine, *Review and Assessment of Planetary Protection Policy Development Processes,* The National Academies Press, Washington, D.C., 2018, pp. 101-105.

extraterrestrial bodies whose environmental conditions are consistent with the maintenance of life.[62,63] However, as of today, no categorization has taken place for the martian moons. Phobos and Diemos are a special case not because of what they are but because of where they are located.

The current categorization of planetary protection target bodies for Category V (i.e., all sample return) missions is as follows:

- *Restricted Earth return*—Mars, Europa, Enceladus, and other to-be-determined (TBD) bodies[64]
- *Unrestricted Earth return*—Venus, Moon, and other TBD bodies

The close proximity of Phobos and Deimos to Mars greatly complicates the planetary protection calculations because major impacts on Mars can scatter martian material throughout cismartian space. Some of the ejected martian material will end up on Phobos and Deimos. A sample return from the martian moons could be effectively a Mars sample return mission, and such missions are classified as restricted Earth return.

Samples classified as restricted Earth return are subject to stringent pre- and post-flight requirements. Current COSPAR policy mandates, in part, the following:[65]

> For all other Category V missions, in a subcategory defined as "restricted Earth return," the highest degree of concern is expressed by the absolute prohibition of destructive impact upon return, the need for containment throughout the return phase of all returned hardware which directly contacted the target body or unsterilized material from the body, and the need for containment of any unsterilized sample collected and returned to Earth. Post-mission, there is a need to conduct timely analyses of any unsterilized sample collected and returned to Earth, under strict containment, and using the most sensitive techniques. If any sign of the existence of a nonterrestrial replicating entity is found, the returned sample must remain contained unless treated by an effective sterilizing procedure. Category V concerns are reflected in requirements that encompass those of Category IV plus a continuing monitoring of project activities, studies and research (i.e., in sterilization procedures and containment techniques).

Implementing the preceding requirements and more is complicated, time consuming, and expensive. The MMX mission builds heavily on the heritage of JAXA's successful Hayabusa 1 and ongoing Hayabusa 2 asteroid sample return missions. The asteroids visited by the Hayabusa missions were both categorized as unrestricted Earth return. The associated contamination avoidance and containment requirements were minimal to none, other than protecting the samples from being contaminated by Earth. Making MMX compliant with restricted Earth return requirements would mean costly redesign of the spacecraft so that samples were strictly contained. It would also have to be designed to break the chain of contact between the moon sampled and any uncontained portions of the spacecraft actually returned to Earth. In addition, a receiving facility would need to be constructed that is capable of both protecting Earth from the samples and the samples from Earth.

The COSPAR requirements outlined above do not specify particular sterilization or containment protocols. Other groups have looked at the specifics and have recommended that protecting Earth from the samples requires that the receiving facility operate at a standard equivalent to a biosafety level (BSL)-4 biological containment

[62] The status of sample return missions from small solar system bodies, for example, is determined by a test consisting of six questions, as follows. Does the preponderance of scientific evidence indicate (1) That there was never liquid water in or on the target body? (2) That metabolically useful energy sources were never present? (3) That there was never sufficient organic matter (or CO_2 or carbonates and an appropriate source of reducing equivalents) in or on the target body to support life? (4) That subsequent to the disappearance of liquid water, the target body has been subjected to extreme temperatures (i.e., 433 K)? (5) That there is or was sufficient radiation for biological sterilization of terrestrial life forms? (6) That there has been a natural influx to Earth—for example, via meteorites, of material equivalent to a sample returned from the target body? Returning six "no" or "uncertain" answers requires that the sample return mission be designated restricted Earth.

[63] G. Kminek, C. Conley, V. Hipkin, and H. Yano, 2017, COSPAR Planetary Protection Policy, *Space Research Today* 200:12-25.

[64] With TBD indicating that additional analysis is required.

[65] G. Kminek, C. Conley, V. Hipkin, and H. Yano, 2017, COSPAR Planetary Protection Policy, *Space Research Today* 200:14-15.

laboratory.[66] Similarly, others have suggested a containment criterion of one in a million for the release of any particle ≥10 nm.[67]

The categorization of Phobos and Deimos is the subject of the present report. In order to proceed with the mission as currently planned and remain consistent with current planetary protection practice, JAXA needs to demonstrate that the probability of a single unsterilized particle from Mars, ≥10 nm in diameter, in an uncontained sample returned from Phobos or Deimos is less than 10^{-6}. If the probability is greater than 10^{-6}, then JAXA faces three alternatives: redesign MMX to be consistent with restricted Earth return requirements, cancel the program, or change the requirements.

STUDIES IN SUPPORT OF PLANETARY PROTECTION CLASSIFICATION OF MARTIAN MOONS SAMPLE RETURN

As mentioned in the preceding section, the planetary protection categorization for a sample return mission specifically from the martian moons is still to be decided. With the MMX mission being planned, it became necessary for relevant space agencies to develop a planetary protection policy for the martian moons.

As already mentioned, large meteorite impacts on Mars are expected to eject material from the planet's surface, and some of this ejecta will ultimately be deposited on the martian moons. Sample return missions to the Phobos and Deimos therefore represent opportunities to collect pristine minerals and, potentially, molecular evidence of life transferred from the surface of Mars. Therefore, the potential for martian moons sample return missions to collect unsterilized martian material needs to be investigated.

The present report reviews the results of two such studies, one sponsored by ESA and the other by JAXA.

In 2014, ESA tasked Manish Patel and his team at the Open University to conduct "feasibility studies and tests to determine the sterilization limits for sample return planetary protection measures." The final objective of that study was to evaluate the probability that unsterilized martian material could be naturally transferred to Phobos, and whether that material would be accessible to a Phobos and, by extension, a Deimos sample return mission. The Open University team produced several reports dealing with the various aspects of the material transfer from Mars to the martian moons due to a crater forming impact. The team and the report it produced will henceforth be referred to as, respectively, "the SterLim team" and "the SterLim report."[68]

Additionally, JAXA tasked a multi-institutional review team, led by Kazuhisa Fujita,[69] to assess the microbial contamination probability for sample return from the martian moons. The purpose of the study was to clarify the potential physical processes that can bring about microbial contamination on the surface of martian moons, to obtain a quantitative estimate of the density of microorganisms still surviving in the regolith of the martian moons through several sterilization processes, and to assess microbial contamination probability of samples collected on the surface of the martian moons for future sample return missions from the martian moons. The aforementioned study was presented to this committee and will be referred to as "the JAXA report" in the following sections.[70]

The results of both the SterLim and the JAXA studies were used to help assess the level of planetary protection measures that need to be implemented for a future sample return mission to Phobos and Deimos to mitigate the risk of release of nonterrestrial life into Earth's environment upon delivery.

---

[66] National Research Council, 2002, *The Quarantine and Certification of Martian Samples*, The National Academies Press, Washington, D.C.

[67] These criteria—10 nm and 10^{-6}—were recommended in an ESF study—*Mars Sample Return Backward Contamination-Strategic Advice and Requirements*, Report from the ESF-ESSC Study Group on MSR Planetary Protection Requirements, European Science Foundation, Strasbourg, France, 2011—and have subsequently been endorsed by ESA's Planetary Protection Working Group. These criteria have not yet been officially adopted by COSPAR or NASA.

[68] From the SterLim consortium, the authors of the reports. The consortium included the Open University, Public Health England, Thales Alenia Space, Kallisto Consultancy, and Fluid Gravity Engineering.

[69] The JAXA review team included experts from the Institute of Space and Astronautical Science, the Chiba Institute of Technology, the Tokyo Institute of Technology, and the University of Tokyo and JAXA scientists.

[70] K. Fujita, K. Kurosawa, H. Genda, R. Hyodo, T. Mikouchi, S. Matsuyama, and the Phobos/Deimos Microbial Assessment Team, 2018, "Assessment of Microbial Contamination Probability for Sample Return from Martian Moons," GNG-2018003, http://sites.nationalacademies.org/SSB/CurrentProjects/SSB_181917.

In addition to the reports commissioned above, ESA and JAXA, with NASA also participating (see the Preface), also requested an independent review of these reports. As mentioned in the Preface, the present report is the result of this assessment by a joint European, Japanese, and American team of experts. The outcome and final recommendations of this review process are detailed in the following chapters.

2

Overview and Assessment of the SterLim and JAXA Studies

The SterLim (Sterilisation Limits for Sample Return Planetary Protection Measures) report was presented to the committee at its meeting in 2017. An updated version of the SterLim report and the Japan Aerospace Exploration Agency (JAXA) report were presented at the committee's meeting in 2018.[1,2,3] The following sections are based on the review of the committee members for the various aspects of these reports and their findings. This chapter contains a brief outline of the work and the methodology the two teams followed. The structure of the chapter mirrors the process of microbial transfer, as modeled by the teams (Figure 2.1). This chapter also contains the committee's review of the work presented to them for each step in the process.

The scenario that both teams considered for their respective studies assumed an initial microbial density in the surface of Mars and estimated how much of the microbial community would survive transfer from Mars to its moons. For the purposes of this report, the surface of Mars includes all material capable of being ejected from the planet during a major impact. Thus, the surface includes all material up to a depth of several km. Any martian microbes would pass through several phases of "sterilization" or dilution in their transfer from Mars to its moons. These phases are as follows:

- During the initial impact;
- During the aerodynamic heating when passing through the martian atmosphere;
- During the hypervelocity impact of the material on Phobos or Deimos and their subsequent recirculation and reimpact to the moons' surface;
- From radiation on the surface of the moons; and
- From reformation and "gardening" of the moons' surface from the natural meteoroid flux.

---

[1] M. Patel, V. Pearson, D. Summers, D. Evans, A. Bennet, and P. Truscott, SterLim, 2017, "Sterilisation Limits for Sample Return Planetary Protection Measures 'SterLim': Presentation to the Committee on the Review of Planetary Protection Requirements for Sample Return from Phobos and Deimos," http://sites.nationalacademies.org/cs/groups/ssbsite/documents/webpage/ssb_183902.pdf; referred to here as the "SterLim report."

[2] M. Patel, D. Summers, and D. Evans, 2018, "SterLim Team Presentation," presentation to the committee on September 18.

[3] A. Yamagishi and K. Fujita, 2018, "JAXA's Planetary Protection Activities for Sample Return from Martian Moons," TR 2018-13-02, presentation to the committee on September 18, 2018, http://sites.nationalacademies.org/SSB/CurrentProjects/SSB_181917.

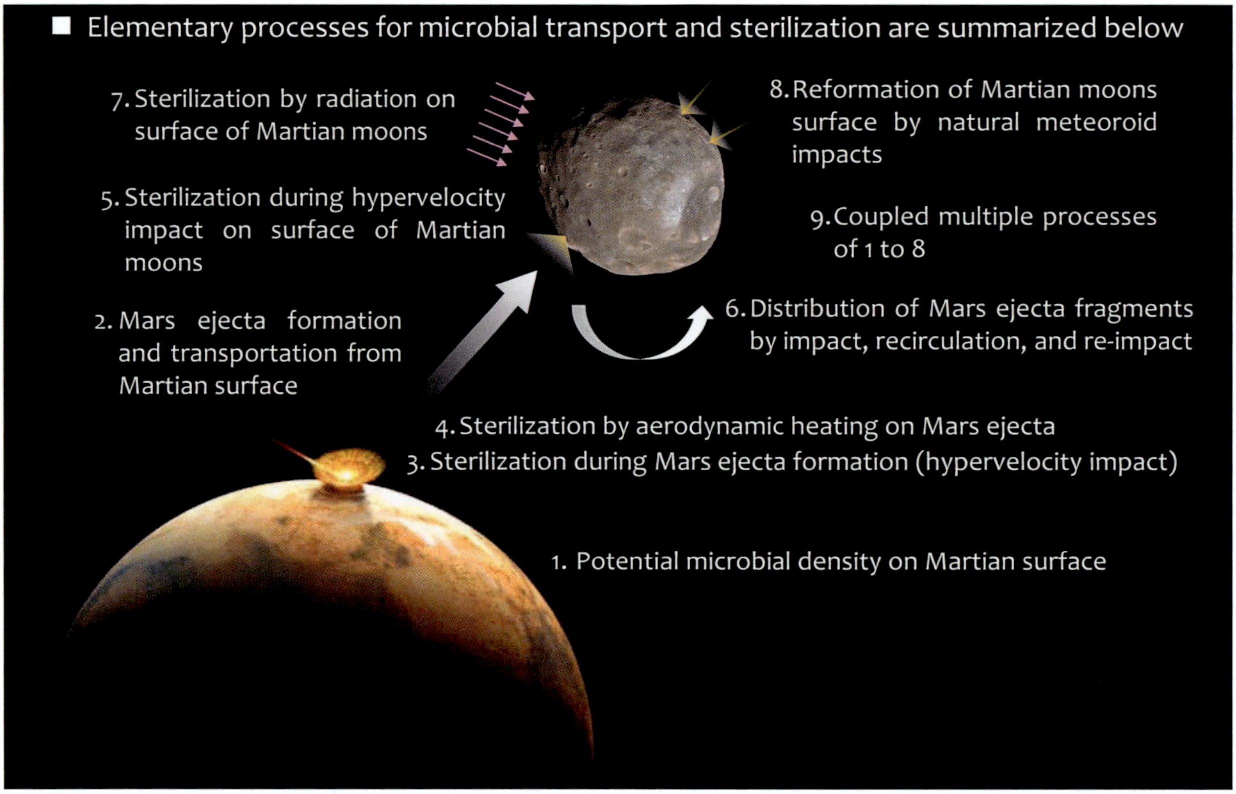

FIGURE 2.1 The various steps of martian material transferred to Phobos (and inferred for Deimos). The SterLim (Sterilisation Limits for Sample Return Planetary Protection Measures) and JAXA (Japan Aerospace Exploration Agency) teams undertook experimental studies or numerical modeling to study each distinct step in the chain from the surface of Mars to that of its moons. The committee organized its report around the various steps outlined here. SOURCE: A. Yamagishi and K. Fujita, 2018, "JAXA's Planetary Protection Activities for Sample Return from Martian Moons," TR 2018-13-02, presentation to the committee, http://sites.nationalacademies.org/SSB/CurrentProjects/SSB_181917, used with permission.

Each phase represents a specific model developed by the SterLim and JAXA teams or certain assumptions made. For example, Step 1 in Figure 2.1 requires that certain assumptions be made about the abundance of microbes in the martian surface material.

The subsequent sections of this chapter discuss each of the first eight steps outlined in Figure 2.1 and describe the modeling activities and assumptions made by both the SterLim and JAXA teams.

POTENTIAL MICROBIAL DENSITY ON THE MARTIAN SURFACE

The potential microbial density on Mars is, of course, a completely unknown parameter. The SterLim study considered the density to be similar to the Atacama Desert and used a range of possible bioloads (Table 2.1) to account for the uncertainties both in Earth measurements and in the potential variability of environments on Mars.

The JAXA report revised SterLim's bioload estimates (see Table 2.1) based on measurements done in Antarctic analogues that are arid and cold (and thus, presumably, closer to a martian environment). The JAXA team concluded that Antarctic microbial density is of the same order or lower than the Atacama Desert. However, due to the lack of any scientific evidence on the baseline value for Mars, both studies have accepted a maximum value based on the Atacama studies, in order to provide a conservative assessment.

TABLE 2.1 Bioload Estimates Used by the SterLim and JAXA Teams Expressed in Colony Forming Units (cfu) per Kilogram of Martian Surface Material

Martian Bioload Estimates Used	SterLim	JAXA
Low	10^5 cfu/kg	Not used
Medium	10^7 cfu/kg	Not used
High	10^{10} cfu/kg	10^8 cfu/kg

NOTE: JAXA, Japan Aerospace Exploration Agency; SterLim, Sterilisation Limits for Sample Return Planetary Protection Measures.

Committee Assessment

Whether or not life exists on Mars is completely unknown. If it does exist, then its biochemical nature is even more of an unknown. If life actually exists on Mars, its molecular underpinning fits somewhere on a continuum with life as it exists on Earth—that is, carbon based, using water as a solvent, employing DNA to transfer genetic information, and employing RNA to control the synthesis of proteins—at one extreme, to organisms whose underlying chemistry is completely different from that of all known life on Earth. Planetary protection calculations are, in a sense, an exercise in due diligence, with researchers ruling out what they can based on what they know. If martian life is assumed to be fundamentally different from all life known on Earth, then researchers have no foundation on which to make planetary protection decisions. But, if researchers assume that life on Mars is basically similar to life on Earth, then they can make some progress. Mars today is cold and arid, so current understanding of life existing in the coldest and most arid regions on Earth may provide some insights into possible life-forms existing on Mars now and in the recent past.

Both the SterLim report and the JAXA report employed the values of the colony forming unit for the cell density of the Atacama Desert. However, the values of the unit are inappropriate as proxies of the cell density because colony forming ability generally depends on species of microbes. Namely, certain species of microbes have already been identified under the growth conditions on given medium agar plates, but others have not. The use of cell density values enumerated by direct cell count or biomolecules-based count (lipid, DNA, others) would have been more appropriate. Connon and coworkers reported that phospholipid fatty acid (PLFA) concentrations ranged from 2×10^5 to 7×10^6 cell equivalents per gram of soil, and Lester et al. reported that PLFA analysis indicated 2.0×10^6 to 1.0×10^7 cell equivalents per gram. Thus, the cell density of the Atacama Desert is likely to be greater than what was used by the SterLim and JAXA teams.[4,5]

Meanwhile, even if the dryness of Atacama is similar to Mars, it is probably true that the Atacama Desert is an energetically favorable environment for life compared with Mars. A close examination of the microbial communities in the soil of the Atacama Desert reveals that the ecosystem is governed by bacterial photosynthesis. In such ecosystems, several microbial species (cyanobacteria) are capable of using solar energy (a huge energy source) to fix carbon and thus form the basis of the food chain for the microbes surrounding the photosynthetic bacteria. Therefore, it would be better to use the values of cell density in the subsurface of the Atacama Desert. This would have minimized the effect of photosynthesis, which is an unlikely source of primary production on the martian surface given the strong ultraviolet flux.[6] And it is probably reasonable, given that the cell density in the subsurface is much lower than that at the surface.

Regarding the Antarctic analogues, low microbial biomass was found by direct microscopic cell counts (1.4 to 5.7×10^6 cells per kg soil) in both the dry and ice-cemented permafrost. In this case, JAXA's calculation (10^8 cells per kg) is sufficiently conservative.

[4] S.A. Connon, E.D. Lester, H.S. Shafaat, D.C. Obenhuber, and A. Ponce, 2007, Bacterial diversity in hyperarid Atacama Desert soils, *Journal of Geophysical Research* 112:G04S17, doi:10.1029/2006JG000311.

[5] E.D. Lester, M. Satomi, and A. Ponce, 2007, Microflora of extreme arid Atacama Desert soils, *Soil Biology and Biochemistry* 39:704-708.

[6] T.C. Onstott, B.L. Ehlmann, H. Sapers, M. Coleman, M. Ivarsson, J.J. Marlow, A. Neubeck, and P. Niles, 2018, "Paleohosted Rock Life on Earth and the Search on Mars: A Review and Strategy for Exploration," https://arxiv.org/abs/1809.08266.

MARS EJECTA FORMATION AND TRANSPORTATION FROM THE MARTIAN SURFACE

The ejection process following an impact on Mars was modeled by the SterLim team using the models developed by Melosh and coworkers in 2011 under a contract from NASA's Office of Planetary Protection.[7] The SterLim team did not consider any initial sterilization caused by the impact. Detailed modeling of Mars ejecta formation and subsequent transport from the martian surface was not part of the SterLim study's statement of work. It was modeled only to the extent required to predict the mass transferred to the moons. The models were transformed to fit the needs of the Monte Carlo approach used by the SterLim team. They were normalized so that the total mass transported to Phobos in 10 million years corresponds to the value predicted by the models developed by Melosh et al. The SterLim models provided only the mass and velocity distributions of the martian ejecta.

The transit to the martian moons was modeled using orbital mechanics. The impact velocity distribution of martian ejecta on Phobos (or, by extension, on Deimos) obtained by the SterLim team corresponds well to the velocity distribution from Melosh's models. In the Monte Carlo approach, impacts on Mars are modeled as stochastic events. This leads to a uniform transfer of mass from Mars to its moons throughout time.

In an addendum of the study, the SterLim team additionally looked at individual cratering events on Mars ("discrete ejections") in general and focused on ejecta that would be generated from Zunil crater in particular. Zunil is a very young (~1-10 Ma), 10 km diameter impact crater on Mars, and is of particular interest to this study because materials from this impact event would by definition have had residence time <10 million years on the moons' surfaces. In this addendum, the same approach was used to model the ejection process as for the original study.

The JAXA team considered five discrete events on Mars that formed craters that are >10 km in diameter and estimated to be younger than 10 million years. Five craters fit these criteria: Mojave, Tooting, McMurdo, Corinto, and Zunil. No other impact events on Mars were considered, although the contamination risk from unrecognized young craters was also investigated. From the dimensions of these five craters, possible combinations of impactor sizes, impact velocities, and impact angles were identified using a Monte Carlo approach and semi-empiric scaling laws. Corresponding numerical simulations were performed, and the ejecta generated by those simulated events propagated into Mars orbit and onto the moons. To reduce the calculation effort, five impact velocities (between 6 and 18 km/s) and six impact angles (between 0 and 75 degrees from the surface normal) were considered. For all craters, the actual location on Mars was included in the trajectory analysis. Specifically, for Zunil crater, the impact direction was restricted between northeast and east in accordance with a 2007 paper by Preblich et al.[8]

Both the SterLim and the JAXA teams reached the following conclusions:

- Large impacts (crater diameter >10 km) dominate the mass deposition on the martian moons.
- Ejecta deposited on the martian moons from craters older than 1 million years can be safely ignored due to the long residence of any microbes on the moon's surfaces and the subsequent sterilization by the space radiation environment. (See "Sterilization by Radiation on Phobos/Deimos Surfaces," later in this chapter.)

Both teams concluded that the last major impact on Mars with sufficient energy to deposit material on Phobos is the 10-km-diameter Zunil crater. Zunil is located in the equatorial Cerberus plains, one of the youngest areas on Mars and a potential target for future landed exploration.[9] Thus, Zunil-derived materials will be present in the surface regolith of Phobos; less so for Deimos, which is more distant from Mars.

To estimate the mass of martian material ejected at speeds sufficient to reach Phobos's orbit, numerical simulations were performed using the smoothed particle hydrodynamics (SPH) approach. The Tillotson equation of state, with a parameter set for granite, was used for both projectile and target. As neither a gravity nor a strength model was used in the numerical simulations, the results were considered dimensionless and thus scalable to any

---

[7] H.J. Melosh, K.C. Howell, L. Chappaz, M. Vaquero, 2011, "Material Transfer from the Surface of Mars to Phobos and Deimos," 22nd AAS/AIAA Space Flight Mechanics Meeting, Univelt Inc., San Diego, Calif.

[8] B.S. Preblich, A.S. McEwen, and D.M. Studer, 2007, Mapping rays and secondary craters from the martian crater Zunil, *Journal of Geophysical Research: Planets* 112(E5), https://doi.org/10.1029/2006JE002817.

[9] See, for example, A. McEwen, P. Lanagan, R. Beyer, L. Keszthelyi, and D. Burr, 2003, "Potential 2003 Landing Sites in the Cerberus Plains, Se Elysium Planitia," presentation at the First Landing Site Workshop for MER 2003, https://www.lpi.usra.edu/meetings/mer2003/pdf/9022.pdf.

projectile size. The choice of granite, rather than the basalt likely to form much of the martian surface, was not explained. The simulations were validated by visual comparison to a single laboratory experiment with a 5 mm diameter polycarbonate projectile impacting a polycarbonate surface at 3.56 km/s with 45-degree impact incidence.

Committee Assessment

The models used by the two teams result in significantly different amounts of predicted mass ejected from Mars's surface. This discrepancy is attributed to the different approaches followed by the two teams. The SterLim team relied on the work from Melosh and coworkers,[10] who used semi-empirical impact crater models.[11] This approach inherently includes most impact angles, but depends on the distribution of velocity of ejected fragment as a function of the mass of the ejected fragments, which is uncertain, particularly when including the impact angle.[12] The JAXA team, in contrast, used the results from numerical simulations specifically performed for its study. These simulations predict that the mass ejected from a 45-degree impact is three to five times the mass ejected from a normal impact (depending on impact velocity),[13] which is a significant increase. As a result, the mass transferred to Phobos is 1.12×10^6 kg according to Melosh et al. (2011) and 5×10^7 kg according to the JAXA team.[14]

The process of fragmentation and ejecta formation during hypervelocity impacts is currently not well understood. Specifically, the dependence of the mass of ejected material on impact angle is not described consistently in the literature. According to experiments, the final crater volume decreases with increasing impact angle (measured from the surface normal).[15,16] In contrast, numerical simulations predict that the mass of ejected material increases with increasing impact angle.[17,18] The results are not directly comparable. Experimental studies analyze the final crater volume (or, equivalently, ejecta mass), whereas numerical studies analyze the ejected mass above a threshold ejection velocity. Possible sources of this discrepancy are manifold, with two being the strength and porosity of the target material. These two latter factors were neglected in the JAXA simulations. It is known that the strength of the target material affects the final crater morphology of oblique impacts,[19] which is also seen when comparing impact experiments on granite against experiments on low-strength materials.[20,21] The influence of material strength on the ejecta formation process in general, and on the formation of fragments fast enough to reach the martian moons in particular, is not well known for these velocities and size scales. For fragments that are ejected at high velocities, and especially for the material transported to the martian moons, the influencing parameters, including material strength and porosity, are not fully constrained.

---

[10] Melosh et al., 2011, "Material Transfer from the Surface of Mars to Phobos and Deimos."

[11] As described in, for example, J.E. Richardson, H.J. Melosh, C.M. Lisse, and B. Carcich, 2007, A ballistics analysis of the deep impact ejecta plume: Determining comet Temple 1's gravity, mass, and density, *Icarus* 190:357-390, https://doi.org/10.1016/j.icarus.2007.08.001.

[12] J.N. Head and H.J. Melosh, 2000, "Launch Velocity Distribution of the Martian Clan Meteorites," Abstract 1937, *Proceedings of the 31st Lunar and Planetary Science Conference*, https://www.lpi.usra.edu/meetings/lpsc2000/pdf/1937.pdf.

[13] See Figure 8-2 in K. Fujita, K. Kurosawa, H. Genda, R. Hyodo, T. Mikouchi, S. Matsuyama, and the Phobos/Deimos Microbial Assessment Team, 2018, "Assessment of Microbial Contamination Probability for Sample Return from Martian Moons," GNG-2018003.

[14] K. Kurosawa, K. Fujita, H. Genda, R. Hyodo, T. Mikouchi, and Phobos/Deimos Microbial Contamination Assessment Team, 2018, "Assessment of Microbial Contamination Probability for Sample Return from Martian Moons," presentation to committee on September 19, http://sites.nationalacademies.org/cs/groups/ssbsite/documents/webpage/ssb_190400.pdf.

[15] M.J. Burchell and L. Whitehorn, 2003, Oblique incidence hypervelocity impacts on rock, *Monthly Notices of the Royal Astronomical Society* 341(1):192-198, https://doi.org/10.1046/j.1365-8711.2003.06385.x.

[16] T. Michikami, A. Hagermann, T. Morota, J. Haruyama, and S. Hasegawa, 2017, Oblique impact cratering experiments in brittle targets: Implications for elliptical craters on the Moon, *Planetary and Space Science* 135: 27-36, https://doi.org/10.1016/j.pss.2016.11.004.

[17] Kurosawa et al., 2018, "Assessment of Phobos Microbial Contamination by Mars Ejecta."

[18] B. Artemieva and B. Ivanov, 2004, Launch of martian meteorites in oblique impacts, *Icarus* 171:84-101.

[19] D. Elbeshausen, K. Wünnemann, and G.S. Collins, 2013, The transition from circular to elliptical impact craters, *Journal of Geophysical Research: Planets* 118(11):2295-2309, https://doi.org/10.1002/2013JE004477.

[20] M.J. Burchell and L. Whitehorn, 2003, Oblique incidence hypervelocity impacts on rock, *Monthly Notices of the Royal Astronomical Society* 341(1):192-198, https://doi.org/10.1046/j.1365-8711.2003.06385.x.

[21] T. Michikami, A. Hagermann, T. Morota, J. Haruyama, and S. Hasegawa, 2017, Oblique impact cratering experiments in brittle targets: Implications for elliptical craters on the Moon, *Planetary and Space Science* 135: 27-36, https://doi.org/10.1016/j.pss.2016.11.004.

Given the different approaches, the models of the SterLim and JAXA teams predict that significantly different masses of material are transported to the martian moons. For Phobos, the SterLim model predicts 1.6×10^6 kg (comparable to the 1.12×10^6 kg from Melosh et al.),[22] while the JAXA model predicts 5×10^7 kg.[23] The specific origin of this discrepancy between the two approaches could not be identified.

STERILIZATION DURING MARS EJECTA FORMATION

The SterLim team did not include any sterilization during Mars ejecta formation in its analysis because such investigations were not requested in its study's statement of work. However, the JAXA team pointed out that considerable heat is generated by shock heating when an interplanetary impactor collides with Mars. The JAXA team performed a numerical simulation of a 45-degree impact at 3.5 km/s using the iSALE shock physics code.[24,25,26] In this manner, it showed that ejecta faster than 3.8 km/s (i.e., that which can reach Phobos's orbit) experience heating above 1000 K when considering internal friction and plastic deformation.[27] Based on heat inactivation tests conducted by the SterLim team, its JAXA counterpart expected that this heating highly sterilizes the ejecta from Mars. In contrast to these impact physics models, the JAXA study did note that some experimental observations of martian meteorites do not show any signatures of shock heating. (See "Earth Inventory of Martian Meteorites," in Chapter 1.) In summary, the JAXA team decided to assume a survival rate of 10 percent during Mars ejecta formation, but noted that this assumption may be too conservative.

Committee Assessment

The numerical simulation performed by the JAXA team suggests that the ejecta accelerated sufficiently to reach the martian moons experience significant heating that enables sterilization.[28] Additionally, the JAXA team reviewed investigations of martian meteorites, some of which show significant shock heating and others showing no signs of shock heating. (See "Earth Inventory of Martian Meteorites," in Chapter 1.) The JAXA team was therefore not confident in the validity of its results. The simulation was performed for a comparatively low impact speed of 3.5 km/s. At such a low speed, the material accelerated to ejection velocity most likely originates from near the contact surface, where plastic deformation (and heating) is most significant. For higher impact speeds, material farther from the contact surface can also reach ejection velocity, presumably with much smaller plastic deformation. Also, shock heating is a highly localized process, and numerical simulations addressing it require very high spatial resolution. The committee was unable to define a survival rate based on the information available. However, the proposed survival rate of 10 percent is a reasonable estimate, albeit one lacking significant experimental evidence.

---

[22] Melosh et al., 2011, "Material Transfer from the Surface of Mars to Phobos and Deimos."

[23] Kurosawa et al., 2018, "Assessment of Phobos Microbial Contamination by Mars Ejecta."

[24] A.A. Amsden, H.M. Ruppel, and C.W. Hirt, 1980, "SALE: A Simplified ALE Computer Program for Fluid Flow at All Speeds," LA-8095, https://inis.iaea.org/collection/NCLCollectionStore/_Public/11/571/11571883.pdf.

[25] B.A. Ivanov, S. Deniem, and G. Neukum, 1997, Implementation of dynamic strength models into 2-D hydrocodes: Applications for atmospheric breakup and impact cratering, *International Journal of Impact Engineering* 20:411.

[26] K. Wünnemann, G.S. Collins, and H.J. Melosh, 2006, A strain-based porosity model for use in hydrocode simulations of impacts and implications for transient crater growth in porous targets, *Icarus* 180(2):514-527, https://doi.org/10.1016/j.icarus.2005.10.013.

[27] See Figure 4-1b of K. Kurosawa and H. Genda, 2018, Effects of friction and plastic deformation in shock-comminuted damaged rocks on impact heating, *Geophysical Research Letters* 45(2):620-626, https://doi.org/10.1002/2017GL076285.

[28] K. Kurosawa and H. Genda, 2018, Effects of friction and plastic deformation in shock-comminuted damaged rocks on impact heating, *Geophysical Research Letters* 45(2):620-626, https://doi.org/10.1002/2017GL076285.

STERILIZATION BY AERODYNAMIC HEATING OF MARS EJECTA

The passage through the martian atmosphere was considered only by the JAXA team and not by the SterLim team. Using computational fluid dynamics simulations, heating and aerodynamic deceleration of single spherical ejecta fragments during the passage through the martian atmosphere were assessed. For this step in the chain, those parts of the fragment were considered sterilized where a temperature above 773 K was maintained for more than 0.5 s.[29] The JAXA team concluded that aerodynamic heating could cause a microbial survival rate below 14 percent for 2 cm size fragments depending on impact angle. For fragments above 4 cm size, more than 50 percent of microbes were predicted to survive. Therefore, sterilization by aerodynamic heating was not considered to be significant in the further analyses because particles less than 10 cm are unlikely to reach Mars orbit due to aerodynamic deceleration.

Committee Assessment

The JAXA model considers only single ejecta fragments passing through the martian atmosphere. The major conclusion from the aerodynamic analysis was that particles smaller than 10 cm are unlikely to reach Mars orbit due to aerodynamic deceleration. However, an impact event creates numerous particles that pass through the atmosphere within a very short time period. Therefore, it is to be expected that much of the ejecta will interact with a disturbed atmosphere.

A mass comparison shows that the ejected mass traveling at high velocities is approximately of the same order of magnitude as the disturbed atmospheric mass. According to McEwen et al.,[30] numerical simulations of the Zunil impact event suggest that 1.5 km^3 of martian rocks were ejected at velocities greater than 1 km/s, which translates to 4×10^9 kg assuming a density of 2.6 g/cm^3. The mass of the atmosphere directly above the crater is roughly 1.3×10^{10} kg (calculated from 600 Pa Mars surface pressure, 3.7 ms^{-2} Mars surface gravity, and 10.1 km crater diameter). Hence, the influence of the ejecta particles on the atmosphere is considered significant. This effect reduces the aerodynamic deceleration and brings into question any subsequent analyses making use of this lower limit on the size of particles capable of escaping the martian atmosphere. However, it is clear, as the JAXA team concluded, that aerodynamic heating of ejecta during passage through the martian atmosphere does not cause any significant sterilization.

STERILIZATION DURING HYPERVELOCITY IMPACT ON PHOBOS/DEIMOS SURFACES

Hypervelocity impact experiments, designed to assess impact survivability for four types of organisms, were carried out by the SterLim team using the Open University's two-stage, light-gas gun. In each experiment, selected organisms (see "Sterilization by Radiation on Phobos/Deimos Surfaces," later in this chapter, for details) were loaded into a hole drilled into the rear face (relative to the velocity vector) of cylindrical projectiles. Experimental design was confined to one projectile material (solid basalt slugs, 3 mm in diameter). The one target material was a Phobos regolith simulant sieved to particle sizes 400 μm and under, with an approximate porosity of 23 percent and bulk density of 1.51 g/cm^3. The target buckets were small, only 70 mm in diameter,[31] to reduce the required regolith mass to ~100 g per experiment. Exploration of sensitivity to target and projectile material choices was not pursued apart from early attempts with concrete targets, which sterilized everything.

The results presented for organism survival at velocities 0.5-1.8 km/s exhibited many orders of magnitude of spread at a given velocity. The SterLim team assessed that no organisms will be killed as a result of an impact at 0.5 km/s, attributing losses at those speeds to flawed collection methods. In a given experiment, the fraction of successfully recovered projectile fragments was very difficult to assess, given compositional similarity between the target and projectile materials chosen and insufficient target confinement. At larger impact speeds, the SterLim team

[29] See Section 5.4 of Fujita et al., 2018, "Assessment of Microbial Contamination Probability for Sample Return from Martian Moons."

[30] A.S. McEwen, B.S. Preblich, E.P. Turtle, N.A. Artemieva, M.P. Golombek, M. Hurst, R.L. Kirk, D.M. Burr, P.R. Christensen, 2005, The rayed crater Zunil and interpretations of small impact craters on Mars, *Icarus* 176(2):351-381, https://doi.org/10.1016/j.icarus.2005.02.009.

[31] See Figures 8 and 9 of Patel et al., 2018, "SterLim Team Presentation."

asserted that the collection methods were more robust and projectile/organism material was not lost. Experiments were carried out only up to 1.8 km/s, while the stated requirements for the SterLim study included velocities up to 4.5 km/s.[32] This requirement was based on the expected velocity distribution of martian material impactors on Phobos and Deimos. The primary diagnostic used for the experiments was the unsterilized fraction of the organisms; planned pyrometer measurements were not successful.

Projectile survival is known to be highly dependent upon impact angle, which affects the peak pressures experienced and the way in which the impactor fails.[33] The SterLim study included vertical shots (90 degrees to surface) and near-horizontal shots with a tilted target bucket (~50 degrees to surface).

The JAXA team performed Monte Carlo simulations to obtain the distribution of collision velocity of Mars rocks onto Phobos and Deimos. It showed that most of the primary ejecta has an impact velocity between ~1-to-20 km/s. The JAXA team used the sterilization data obtained by the SterLim team as input directly into its calculations and did not conduct any additional hypervelocity impact experiments. Unlike the SterLim team, it fit the SterLim sterilization versus impact velocity data directly (without any modeling) using a specific exponential function that it assumed could be applied at the high velocities of the Mars rock impact on the moons.[34]

Committee Assessment

Several issues that were not considered during the SterLim impact experiments and their subsequent analyses may have had significant influence on the results. Since the SterLim results were then used directly by the JAXA team, those errors and uncertainties propagated into the JAXA models. In addition, the JAXA team drew some conclusions from the SterLim results that were very broad generalizations in terms of sterilization from hypervelocity impact, and the committee heard explicitly from the SterLim team that it believed that those generalizations were unfounded.

In each of the SterLim impact experiments, organisms were loaded into a hole drilled into the rear face of cylindrical projectiles. This raised the possibility that high gas temperatures during the launch phase may have contributed to organism sterilization prior to impact because the sabot, into which the projectile is loaded, did not seal off the gas from the projectile. An estimation of the driver gas conditions during launch yields temperatures in excess of 800 K during launch to 2 km/s, which can last for up to 0.8 ms (assuming a 0.1 g projectile, a 0.8 m launch tube, and 1 MPa initial nitrogen fill pressure).[35,36] At lower velocities, the temperatures are considerably lower, but last longer (360 K for 3 ms for a 500 m/s launch, 540 K for 1.6 ms for a 1 km/s launch).

Additionally, as the shock wave hits the rear face of the projectile during impact, the projectile will go into tension, causing fragments to be launched at speeds up to twice the particle velocity in the reference frame of the projectile. Hence, the backside of the projectile is a particularly unfavorable location for organism placement, because of likely loss of biological material, even though it may be perhaps the most favorable location for survival in the laboratory experiment because organisms are separated from the projectile before heat is effectively transferred to the organisms. Thus, it is not clear that the SterLim team's assertion that projectile/organism material was not lost during impact is well supported.

The SterLim team justified ending experiments at 1.8 km/s due to high levels of organism sterilization at that velocity. However, given the large variations in sterilization results, this is not a particularly robust argument. Data at higher velocities would have been useful for validation of the complementary numerical simulations, provided that appropriate diagnostics were employed. Such data would also have been useful in providing better guidance to the JAXA team in terms of sterilization due to hypervelocity impact.

---

[32] See Requirement 190 in Patel et al., 2018, "SterLim Team Presentation."

[33] See, for example, R.T. Daly and P.H. Schultz, 2016, Delivering a projectile component to the vestan regolith, *Icarus* 264:9-19.

[34] See Equation 6.1 in Fujita et al., 2018, "Assessment of Microbial Contamination Probability for Sample Return from Martian Moons."

[35] J. Lukasiewicz, 1967, Constant acceleration flows and applications to high-speed guns, *AIAA Journal* 5(11):1955-1963, https://doi.org/10.2514/3.4346.

[36] G.P.T. Wilenius, P.L. Cowan, and M. Cloutier, 1964, "The Constant Base Pressure Light Gas Gun," in *Proceedings of the 3rd Hypervelocity Techniques Symposium*, Denver Research Institute, Denver, Colo.

The SterLim team claimed that impact angle effects were unimportant, due to no discernible trend between the 90- and 50-degree shots. However, since the variability in the data was very high, even within each impact angle group, this claim was not well supported. Again, problems in experimental design and collection methods cast doubt on the conclusion about impact angle. As the impact angle probability distribution is sinusoidal and centered around 45 degrees for natural planetary impacts, near-vertical impacts are exceedingly rare.[37,38]

Understanding the role of impact angle in organism survival is an important piece of the problem. The SterLim team actually illustrated this point, to some extent, with a handful of three-dimensional (3D) simulations presented in the numerical impact study: A significantly larger fraction of the projectile was not heated to the sterilization limits in the 30-, 20-, and 10-degree simulations, as compared with a two-dimensional (2D), 90-degree impact simulation.

Throughout the study, it was claimed that a regolith "bounce back" effect would result in nearly all projectile material remaining at the surface. There is no physical justification for this assertion, however. The SterLim team set the depth deposited to be one radius of the impactors in its modeling, although experimentalists recognize that fragments can be buried to depths of more than one impactor radius in hypervelocity impact experiments using porous particulate targets (regolith-like).[39] Depth of burial for projectile material is a significant factor to consider, as it could preserve organisms from sterilizing radiation on the surface of Phobos or Deimos.

Overall, the experimental hypervelocity impact data generated during the SterLim study is limited with respect to the large spectrum of possible impact conditions on martian moons, could be biased, and is not conclusive. Given the small footprint of the data within the vast parameter space, extrapolations drawn from this data seemed to the committee to be ill-advised. The committee notes that SterLim's impact data was then used to calibrate the exponential function that the JAXA group used to estimate the likely sterilization due to impact.[40]

DISTRIBUTION OF MARS EJECTA FRAGMENTS BY IMPACTS, RECIRCULATION, AND REIMPACT

The SterLim team assumed that spatially uniform martian ejecta impact with the moons with a random impact angle.[41] Part of the martian materials is deposited on the surface of the moons, and the rest is ejected from the satellites. A part of the ejected materials becomes a "debris cloud" orbiting Mars. Based on the numerical simulation of impact,[42] it was assumed that the impacting material with angle between vertical (0 degree from the surface normal) and 45 degrees is fully deposited on the moons. It was also assumed that the deposited fraction linearly decreases with the angle for highly oblique incidence (>45 degrees). As a result, the percentage deposited on the Phobos surface at the first collision will be 78 percent and on the Deimos surface will be 82 percent. The burial depth of the martian material left on the surface depends on the properties of the moons' regolith and the size of the impactor. Burial depths of one to a few times the radius of the impactor—that is, fragments of martian ejecta—were expected, and the SterLim team reported burial depths of one ejecta-fragment radius in its report. Most of the martian materials were expected to be deposited near the surface.

The SterLim team assumed that the normal velocity component of the ejecta from the moon is 40 percent of the normal component of the impact velocity and pointing away from the moon. The tangential component

---

[37] G.K. Gilbert, The Moon's Face: A Study of the Origin of its Features, 1893 presidential address to the Philosophical Society of Washington and reported in *Science* 21(539):305-307, http://science.sciencemag.org/content/sci/ns-21/539/305.4.full.pdf.

[38] E.M. Shoemaker, 1962, "Interpretation of Lunar Craters," pp. 283-359 in *Physics and Astronomy of the Moon* (Z. Kopal, ed.), Academic Press, London.

[39] See Section 7.2.4 in Thales Alenia Space, 2018, "Evaluation of the Level of Assurance that No Unsterilized Martian Material Naturally Transferred to Phobos (and Deimos) Is Accessible to a Phobos (and Deimos) Sample Return Mission," SterLim-Ph2-TAS-TN21, June 9, http://sites.nationalacademies.org/cs/groups/ssbsite/documents/webpage/ssb_190403.pdf; referred to as "SterLim TN-21."

[40] See Equation 6-1 in Fujita et al., 2018, "Assessment of Microbial Contamination Probability for Sample Return from Martian Moons."

[41] See Thales Alenia Space, 2018, "Sterilisation Limits for Sample Return Planetary Protection Measures: TN 19 Description of the Sterilisation Statistical Model," SterLim-Ph2-TAS-TN19, June 9, http://sites.nationalacademies.org/cs/groups/ssbsite/documents/webpage/ssb_190404.pdf, and SterLim TN-21.

[42] See Fluid Gravity Engineering Ltd., 2017, "Report on the Hypervelocity Impact Modelling," SterLim-PH2-FGE-TN-18, May 10, http://sites.nationalacademies.org/SSB/CurrentProjects/SSB_181917#presentations.

was assumed to be from 76 percent (45-degree impact to the surface) to 100 percent (grazing impact). The team considered that the material going into the cloud undergoes little impact heating during impact and excavation, and so will not be sterilized. It assumed that the ejected material remains at the original size, and therefore the orbital motion is hardly affected by solar irradiation. Most of the ejecta escaped from the Mars system, but half of the remaining (i.e., cloud component) reimpacted with the moons and the other half impacted with Mars. The team assumed that the radiation sterilization is negligible for the cloud component. It showed that the reimpacting velocity distribution to Phobos/Deimos shifts to a lower velocity than the velocity distribution of the direct component from Mars, and sterilization of the cloud component due to impact heating during the second collision is less likely to occur than direct impact component—that is, those deposited at the first collision.

As previously described (see "Mars Ejecta Formation and Transport from the Martian Surface," earlier in this chapter), the JAXA team estimated the amount of martian ejecta with a velocity that could reach the moons by SPH simulations, by changing the impactor's angle and velocity toward Mars.[43] Based on this SPH modeling, the JAXA team estimated the impact velocity and angle of the Mars ejecta onto Phobos and Deimos by the Monte Carlo method. Since the size distribution of Mars ejecta colliding with the moons is unknown, the number of martian rocks colliding at Phobos and Deimos was estimated from the total mass of the ejecta while assuming that the martian rocks were all 10 cm diameter spheres.

The JAXA team estimated the diameter of the "Mars-rock-crater" on the moons using pi-group scaling.[44] Since the stress experienced by martian rocks when impacting onto the regolith of the moon's surface exceeds the estimated compressive strength of the rocks, the team expected that martian rocks will be broken into pieces at the time of impact to form a lens-shape region at the bottom of the resulting crater.[45] A significant fraction of the impactor fragments (i.e., martian rock fragments) is ejected from the regolith surface depending on the collision angle and velocity of martian rocks to the moon's surface.[46] In the case of ejecta from Zunil crater, the percentage of impactor fragments remaining in craters was estimated as 22 and 29 percent for Phobos and Deimos, respectively. A 10 cm diameter martian rock forms a crater with a diameter of ~10 m and a lens thickness of 1 m.

Part of the ejected fragments from the moon's surface will again fall on the surface of the moons. The JAXA team reported that the average thickness of the reaccumulated material will be 30 microns for Phobos and 1 micron for Deimos as a result of the Zunil impact, for example. The team assumed that the thickness of the reaccumulated layer ("global thin layer") is 0.1 mm and expected that it is quickly sterilized by radiation.

Committee Assessment

The SterLim and JAXA teams obtained results for the fraction of martian ejecta deposited on Phobos or Deimos. SterLim expected that about 80 percent is deposited at the first impact to the moons, whereas the JAXA team expected that about 70-80 percent is ejected. The estimations of the two teams are based on the different and limited experimental data (the SterLim team used its own data, and the JAXA team used literature data). Therefore, a factor of uncertainty remains in the fraction deposited at the first impact.

To be conservative, the depth of deposition (one radius of Mars rocks) assumed by the SterLim team needs a modification. The depth of deposition in the JAXA team's model is conservative. The JAXA team's calculations showed that some 80 percent of the fragments of Mars rocks are distributed in a global layer some 0.1 mm thick. Whereas, the remaining 20 percent is mixed with regolith and confined to 1 m thick layers ("collapsed lens") in the bottoms of the craters formed by the impacting martian material.

The model of the JAXA team assumes severe comminution of impactors during the penetration or the ejection. It is true that the dynamic ram pressure exceeds the compressive strength of intact basaltic rocks and Mars rocks would be fragmented. However, this does not mean that the rocks would be fully pulverized into dust (<0.1 mm), but possibly into mm or larger size. Laboratory experiments have shown that the degree of fragmentation depends

[43] See Section 8 of Fujita et al., 2018, "Assessment of Microbial Contamination Probability for Sample Return from Martian Moons."

[44] Ibid.

[45] See the "collapsed lens" in Figure 9-2 of Fujita et al., 2018, "Assessment of Microbial Contamination Probability for Sample Return from Martian Moons."

[46] R.T. Daly and P.H. Schultz, 2016, Delivering a projectile component to the vestan regolith, *Icarus* 264:9-19.

on the ratio of the compressive (or tensile) strength of rocks and the applied stress.[47,48] The size of the largest fragment is highly dependent on the size of the Mars rock and the strength of the Mars rocks that are damaged during their acceleration as they are ejected from the surface of Mars.

STERILIZATION BY RADIATION ON PHOBOS/DEIMOS SURFACES

The SterLim report investigated the killing effects of protons, alpha particles, and gamma rays on test organisms. Ultraviolet light (e.g., UVC) and particle irradiations (e.g., charged particles such as protons and alpha particles) were not factored in microbial survival on Phobos and Deimos, because these forms of radiation are blocked by the top few millimeters of the moons' regolith. These organisms included the extremely radiation resistant bacterium *Deinococcus radiodurans*, the biomedical indicator strains *Brevundimonas diminuta* and *Bacillus atrophaeus*, and an ssRNA virus MS2 coliphage. The representative organisms were irradiated with the objective of determining the radiation conditions required to ensure that sterilization limits are achieved.

Anaerobic, desiccated, and deeply frozen preconditions for the survival of life on Mars before transfer to Phobos and Deimos can be simulated in the laboratory. It follows that the SterLim team was tasked to simulate and demonstrate the level of biological inactivation of materials transferred from Mars to Phobos and Deimos. In an attempt to model better the effects of radiation on possible extant life forms on the surfaces of Mars and its moons, the effects of ionizing radiation (gamma radiation) were studied on desiccated cells and viruses under deeply frozen anaerobic conditions.

Specifically, the SterLim team applied a fixed number of bacteria and viruses directly onto aluminum slides, which were dried, frozen (193 K) and stored under nitrogen gas, and then exposed to acute doses (0-50 kGy) of gamma radiation. Finally, colony- and plaque-forming unit assays for bacteria and viruses, respectively, were used to gauge survival on the irradiated slides. Special emphasis was placed on *D. radiodurans*, the most characterized model for extreme radiation resistance.[49,50,51] *D. radiodurans* is capable of surviving 16 kGy under aqueous conditions, 10-times greater than baker's yeast (*Saccharomyces cerevisiae*), 20-times greater than *Escherichia coli* bacteria, and 3,000-times greater than human cells.[52] In comparison, viruses are generally the life-form most resistant to radiation.[53] Microorganisms were held in their original growth medium, then transferred to the surfaces of sterilized slides, then dried under ambient atmospheric conditions, and finally irradiated.

The results of the irradiation experiments undertaken by the SterLim team are summarized as follows: For the test organisms, the survival metrics of bacteria and viruses exposed to ionizing radiation in the form of gamma rays under *desiccated and deeply frozen conditions* were slightly greater, both in form and scale, to survival metrics reported by others for gamma irradiations performed under *aqueous conditions at 273 K*.[54]

---

[47] Y. Takagi, H. Mizutani, and S.-I. Kawakami, 1984, Impact fragmentation experiments of basalts and pyrophyllites, *Icarus* 59:462-477.

[48] H. Nagaoka, S. Takasawa, A.M. Nakamura, and K. Sangen, 2013, Degree of impactor fragmentation under collision with a regolith surface—Laboratory impact experiments of rock projectiles, *Meteoritics and Planetary Science* 49:69-79.

[49] M.J. Daly, L. Ouyang, P. Fuchs, and K.W. Minton, 1994, In vivo damage and recA-dependent repair of plasmid and chromosomal DNA in the radio-resistant bacterium *Deinococcus radiodurans*, *Journal of Bacteriology* 176:3508-3517.

[50] K.S. Makarova, L. Aravind, Y.I. Wolf, R.L. Tatusov, K.W. Minton, E.V. Koonin, and M.J. Daly, 2001, Genome of the extremely radiation resistant bacterium *Deinococcus radiodurans* viewed from the perspective of comparative genomics, *Microbiology and Molecular Biology Reviews* 65:44-79.

[51] M.J. Daly, E.K. Gaidamakova, V.Y. Matrosova, A. Vasilenko, M. Zhai, A. Venkateswaran, M. Hess, M.V. Omelchenko, H.M. Kostandarithes, K.S. Makarova, L.P. Wackett, J.K. Fredrickson, and D. Ghosal, 2004, Accumulation of Mn(II) in *Deinococcus radiodurans* facilitates gamma-radiation resistance, *Science* 306:1025-1028.

[52] A. Sharma, E.K. Gaidamakova, O.E. Grichenko, V.Y. Matrosova, V. Hoeke, P. Klimenkova, I.H. Conze, et al., 2017, Across the tree of life, radiation resistance is governed by antioxidant Mn^{2+}, gauged by paramagnetic resonance, *Proceedings of the National Academy of Sciences U.S.A.* 114:E9253-E9260.

[53] E.K. Gaidamakova, I.A. Myles, D.P. McDaniel, C.J. Fowler, P.A. Valdez, S. Naik, M. Gayen, P. Gupta, A. Sharma, P.J. Glass, R.K. Maheshwari, S.K. Datta, and M.J. Daly, 2012, Preserving immunogenicity of lethally irradiated viral and bacterial vaccine epitopes using a radio-protective Mn(2+)-peptide complex from Deinococcus, *Cell Host Microbe* 12(1):117-124.

[54] A. Sharma, E.K. Gaidamakova, O.E. Grichenko, V.Y. Matrosova, V. Hoeke, P. Klimenkova, I.H. Conze, et al., 2017, Across the tree of life, radiation resistance is governed by antioxidant Mn^{2+}, gauged by paramagnetic resonance, *Proceedings of the National Academy of Sciences U.S.A.* 114:E9253-E9260.

The JAXA report used the SterLim data for its analysis, averaging the survival rate for specific depth, in order to facilitate the modeling of sample collection with a coring approach, as planned by MMX (see Figure 1.4).

Committee Assessment

The SterLim team's irradiation setup included placing samples of bacteria and viruses directly onto the metal surfaces of aluminum strips. However, aluminum exposed to gamma radiation causes backscattering of electrons at surfaces and forms reactive oxides.[55,56] This increases the proximal dose rates and damage to cells and viruses in contact with the metal during irradiation. The radiation resistance of desiccated *Deinococcus* bacteria would be expected to be higher if irradiated on a chemically inert surface such as polystyrene. Moreover, aluminum metal-cell interactions are not anticipated on the surfaces of Mars or its moons.[57] Overall, the use of aluminum as a simulant environment for irradiation on Mars/Phobos is problematic.

Published desiccation protocols for bacteria including *Deinococcus spp.* involve transferring cells to inert surfaces, then drying in sealed chambers containing desiccants (e.g., Drierite). Polystyrene surfaces prevent secondary reactions with biosamples, and Drierite accelerates evaporation, lessening cell growth during the desiccation process itself, which typically lasts one week.[58] In the absence of hermetically sealed desiccation chambers, the SterLim bacteria might have been washed and resuspended in phosphate buffer prior to desiccation—to prevent continued growth, which can cause large standard deviations in desiccation survival curves.[59]

In subsequent discussions, the committee refers to the higher gamma radiation survival values reported for deinococcal cells irradiated as deeply frozen aqueous preparations.

The Radiation Environment on the Martian Moons

A critical issue upon which support or not for the presence of preserved life on Phobos and Deimos transferred from Mars is radiation—mainly, ionizing forms. If a radiation-resistant organism similar to *D. radiodurans* were to be maintained as fully cryptobiotic in a desiccated habitat in near-surface martian environments, the SterLim team claims that the theoretical accumulated dose maximally allowed without overwhelming an ecological population might approach 20 kGy. At the surface of Mars, and 10 cm below the surface, the background dose rates for ionizing radiation range from 76-96 mGy/year;[60] however, below several meters, the background radiation dose rate is significantly less. Therefore, the theoretical maximal survival dose of deeply frozen, desiccated *D. radiodurans* will be reached in Mars surface environments (i.e., 0-10 cm) in approximately 200,000 to 260,000 years. In contrast, ionizing radiation survival of deeply frozen, aqueous preparations of *D. radiodurans* is approximately four-times greater than when desiccated.[61,62] It follows, in deeply frozen, aqueous conditions on the planet, that the theoretical ionizing radiation maximal dose (~80 kGy) in martian near-surface environments would be reached in approximately 1 million years (i.e., 80,000/0.076).

---

[55] M. Ravikumar, R. Ravichandran, S. Sathiyan, S. Sudhakar Supe, 2004, Backscattered dose perturbation effects at metallic interfaces irradiated by high-energy x- and gamma ray therapeutic beams, *Strahlentherapie Onkologie* 180:173-178.

[56] K. Kanjana, P. Ampornrat, and J. Channuie, 2017, Gamma-radiation-induced corrosion of aluminum alloy: low dose effect, *Journal of Physics: Conference Series* 860:012041.

[57] V.M. Dekov, V. Arnaudov, F. Munnik, T.B. Boycheva, and S. Fiore, 2009, Native aluminum: Does it exist? *American Mineralogist* 94:1283-1286.

[58] J.K. Fredrickson, S.W. Li, E.K. Gaidamakova, V.Y. Matrosova, M. Zhai, H.M. Sulloway, J.C. Scholten, M.G. Brown, D.L. Balkwill, and M.J. Daly, 2008, Protein oxidation: Key to bacterial desiccation resistance, *ISME Journal* 2:393-403.

[59] Ibid.

[60] See Table 3 in D.M. Hassler, C. Zeitlin, R.F. Wimmer-Schweingruber, B. Ehresmann, S. Rafkin, J.L. Eigenbrode, D.E. Brinza, et al., 2014, Mars' surface radiation environment measured with the Mars Science Laboratory's Curiosity Rover, *Science* 343(6169):1244797.

[61] R.C. Richmond, R. Sridhar, and M.J. Daly, 1999, Physicochemical survival pattern for the radiophile *Deinococcus radiodurans*: A polyextremophile model for life on Mars, *SPIE* 3755:210-222.

[62] L.R. Dartnell, S.J. Hunter, K.V. Lovell, A.J. Coates, and J.M. Ward, 2010, Low-temperature ionizing radiation resistance of *Deinococcus radiodurans* and Antarctic Dry Valley bacteria, *Astrobiology* 10:717-732.

For microorganisms *directly* on the surface of Mars or its moons, ultraviolet light and charged particles govern survival.[63] However, just below the surface (>5 mm), extending several meters down, much more penetrating ionizing radiation predominates.[64] For all cosmic radiation forms, the martian atmosphere attenuates dose rates by a factor of two.[65] However, because Phobos and Deimos lack atmospheres, the theoretical maximal ionizing radiation dose would be reached twice as fast, in approximately 125,000 years when desiccated at their surfaces. This sterilization value can be accepted with a high degree of certainty, as there would be no possibility of a reconstituting event, such as hydration, to revive a microbial population on Phobos or Deimos. Moreover, the temperature of the upper few cm of Phobos and Deimos regolith intermittently reach or exceed 273 K, the freezing point of water (see Table 1.1), caused by solar heating. In vacuo, this diurnal temperature cycling causes desiccation, which is bactericidal to even the most radiation-resistant microbes in a matter of months at temperatures above the freezing point of water.[66] It is a significant omission that neither the SterLim nor the JAXA teams accounted for the desiccating effect of diurnal temperature changes. Chapter 3 will return to this again because it will be an important factor in determining whether or not samples from the martian moons are designated restricted or unrestricted Earth return.

PHOBOS/DEIMOS SURFACE REFORMATION BY NATURAL METEOROID IMPACTS

The SterLim team noted that the turnover due to meteorite collision on the lunar surface (i.e., impact gardening) occurred with a confidence of 50 percent down to a depth of ~2 cm at least once every 10 million years. Thus, given that the meteorite flux at the martian system is even lower that at the Moon, the SterLim team concluded that the effect of impact gardening can be ignored.

The JAXA team showed that the collision of martian rocks forms a global thin layer (<0.1 mm) of materials contaminated with martian rock fragments on the surface of the moons, as previously mentioned. (See "Distribution of Mars Ejecta Fragments by Impacts, Recirculation, and Reimpact," earlier in this chapter.) This layer is radiation (UVC and ionizing forms) sterilized in a short time (i.e., a few thousand years). However, if part of the layer is covered by an ejecta blanket due to a collision from interplanetary space, the blanket acts as a radiation shield. From the size-frequency distribution of martian craters,[67] the size-frequency distribution of the interplanetary impactors to the Mars system was estimated using the pi-group, crater scaling law. Then the impact flux on the satellites was estimated using the cross-sectional area ratio of the satellites and Mars. The thickness of ejecta blanket as a function of the distance from the impact point was obtained based on the pi scaling. The fraction of the shielded area with the blanket of thickness 3 mm or more (the area for which radiation sterilization was not completed within 100,000 years) was found to be about 0.1 percent.[68]

Committee Assessment

The final factor in the chain of potential sterilizing process is the reformation of Phobos's or Deimos's surface by the natural fall of meteoroids from interplanetary space. However, this effect is insignificant due to the low flux of impactors.

In addition to bombardment by meteoroids, the surface of Phobos and Deimos can be physically weathered by thermal fatigue.[69] This mechanism breaks down rocks into smaller pieces, thus exposing new surface area and

[63] D. Matthiä, D.M. Hassler, W. de Wet, B. Ehresmann, A. Firan, J. Flores-McLaughlin, J. Guo, et al., 2017, The radiation environment on the surface of Mars: Summary of model calculations and comparison to RAD data, *Life Sciences in Space Research* 14:18-28.

[64] D.M. Hassler, J.W. Norbury, and G. Reitz, 2017, Mars Science Laboratory radiation assessment detector (MSL/RAD) modeling workshop proceedings, *Life Sciences in Space Research* 14:1-2.

[65] D.M. Hassler, C. Zeitlin, R.F. Wimmer-Schweingruber, B. Ehresmann, S. Rafkin, J.L. Eigenbrode, D.E. Brinza, et al., 2014, Mars' surface radiation environment measured with the Mars Science Laboratory's Curiosity Rover, *Science* 343(6169):1244797.

[66] J.K. Fredrickson, S.W. Li, E.K. Gaidamakova, V.Y. Matrosova, M. Zhai, H.M. Sulloway, J.C. Scholten, M.G. Brown, D.L. Balkwill, and M.J. Daly, 2008, Protein oxidation: Key to bacterial desiccation resistance, *ISME Journal* 2:393-403.

[67] See, for example, W.K. Hartmann, 2005, Martian cratering 8: Isochron refinement and the chronology of Mars, *Icarus* 174(2):294-320.

[68] See Table 7.2 in Fujita et al., 2018, "Assessment of Microbial Contamination Probability for Sample Return from Martian Moons."

[69] M. Delbo, G. Libourel, J. Wilkerson, N. Murdoch, P. Michel, K.T. Ramesh, C. Ganino, C. Verati, and S. Marchi, 2014, Thermal fatigue as the origin of regolith on small asteroids, *Nature* 508:233-236.

making fresh regolith. This thermal fragmentation is predicted to occur at a rate orders of magnitude faster than fragmentation due to micrometeoroid impact. Although the effect was not considered by either the SterLim or the JAXA team, it could significantly enhance the rate at which any biological materials present are exposed and degraded by radiation.

3

Responses to the Statement of Task and Recommendations

Chapter 2 presented an overview of the work of the SterLim and JAXA (Japan Aerospace Exploration Agency) teams, together with a review of the models and assumptions used therein. This chapter investigates some additional arguments regarding planetary protection requirements for a sample return mission from the martian moons. The organization of this chapter is designed to provide the committee's answers to the questions posed in the statement of task (see the Preface).

TASK 1—REVIEW OF CURRENT UNDERSTANDING

The committee's first task was as follows:

> Review, in the context of current understanding of conditions relevant to inactivation of carbon-based life, recent theoretical, experimental, and modeling research on the environments and physical conditions encountered by Mars ejecta during the following processes:
> a. Excavation from the martian surface via crater-forming events;
> b. While in transit through cismartian space;
> c. During deposition on Phobos or Deimos; and
> d. After deposition on Phobos or Deimos.

In this context, the committee reviewed the work of the SterLim and JAXA teams. Subsequent sections of this chapter summarize the methodology of the two teams; the issues identified by the committee were discussed in Chapter 2.

Potential Microbial Density on the Martian Surface

The SterLim team assumed the same microbial density as soils from the Atacama Desert. The JAXA team used a similar number with a slight correction.

Finding: The committee finds that if life exists on Mars, its cell density and even its biochemical nature, is unknown. Therefore, the cell-density estimate employed by the SterLim and JAXA teams is, as is appropriate for a planetary protection calculation, a very conservative estimate based on current understanding of life as it exists in Mars-like extreme environments on Earth.

Mars Ejecta Formation and Transportation from the Martian Surface

SterLim models were based on the 2011 report by Melosh et al.[1] The JAXA team used smoothed particle hydrodynamics (SPH) computations newly conducted for statistical analysis. The models of the SterLim and the JAXA teams predict significantly different amounts of mass transported to the martian moons. For Phobos, the SterLim model predicts 1.6×10^6 kg (comparable to the 1.1×10^6 kg from the 2011 report from Melosh et al.), while the JAXA model predicts 5×10^7 kg.

Finding: The committee cannot identify why there is a significant discrepancy in the amount of material transported to the martian moons as determined by the SterLim and JAXA teams. Nevertheless, these uncertainties represent, in some sense, the current state of the art.

Sterilization During Mars Ejecta Formation

The SterLim team did not use a specific model (i.e., the microbial survival rate was 100 percent). The JAXA team added a sterilization model during meteoroid impact according to a numerical simulation.

Finding: The consensus in the committee is that shock heating is a highly localized process. When trying to resolve this adequately in numerical simulations, very high spatial resolutions are required.

Finding: The committee was unable to define a survival rate based on the information available. However, the proposed survival rate of 10 percent is a reasonable estimate, albeit one lacking significant experimental evidence.

Sterilization by Aerodynamic Heating of Mars Ejecta

The SterLim team did not include such sterilization. The JAXA team conducted thermal analysis of Mars ejecta along various trajectories.

Finding: The committee finds that the JAXA team's conclusion that particles smaller than 10 cm do not escape the martian atmosphere is not well supported, and that subsequent analyses relying on this limit be treated with care.

Finding: The committee supports the JAXA team's conclusion that aerodynamic heating of ejecta during passage through the martian atmosphere does not cause any significant sterilization.

Sterilization During Hypervelocity Impact on Phobos/Deimos Surfaces

The SterLim team calculated a microbe survival rate ~ 0.1 for velocity <2 km/s, based on impact experiments. The JAXA team reworked the impact sterilization model using SPH and trajectory analysis, resulting in a survival rate of only 10^{-4}.

Finding: The committee finds that the experimental hypervelocity impact data generated during the SterLim study are limited with respect to the large spectrum of possible impact conditions on the martian

[1] H.J. Melosh, K.C. Howell, L. Chappaz, and M. Vaquero, 2011, "Material Transfer from the Surface of Mars to Phobos and Deimos," 22nd AAS/AIAA Space Flight Mechanics Meeting, Univelt Inc., San Diego, Calif.

moons, could be biased, and are not conclusive. Given the small footprint of the data within the vast parameter space, extrapolations drawn from the experimental data currently available seemed ill-advised. The committee notes that SterLim's impact data were then used to calibrate the exponential function used by the JAXA group to estimate and extrapolate the likely sterilization due to impact.

Distribution of Mars Ejecta Fragments by Impact, Recirculation, and Reimpact

The SterLim team assumed a homogeneous deposition by averaging the incoming flux. The JAXA team took account of crater formation by Mars ejecta with retention and scattering of Mars ejecta fragments.

Finding: The committee finds that the estimations of the two teams were based on the different and limited experimental data (the SterLim team used its own data, and the JAXA team used literature data). Therefore, a factor of uncertainty remains in the fraction deposited at the first impact.

Sterilization by Radiation on Phobos/Deimos Surfaces

The SterLim team used a sterilization model based on experimental data. The JAXA team used a similar model but averaged the microbial survival rate for its specific depth by integrating the survival rate in the depth direction and then dividing by the depth. This enables the JAXA team to account for the fact that the microbial density decreases toward the exposed surface (the averaged density for each depth is smaller than the local density at each depth, but the sampling operation gathers material along the full depth path).

Finding: The committee finds that the use of aluminum, rather than chemically inert surfaces, as a simulant environment for irradiation on Mars/Phobos is problematic. In addition, the samples were irradiated in a frozen state, whereas the surface temperatures on the surfaces of the martian moons is frequently above the freezing point of water.

Finding: The committee finds that diurnal temperature cycling is an extremely significant factor in determining the survival of martian organisms deposited on the surfaces of Phobos or Deimos. Desiccation is bactericidal to even the most radiation-resistant microbes in a matter of months.

Phobos/Deimos Surface Reformation by Natural Meteoroid Impacts

The SterLim team expected this effect to be low and did not take it into account in its calculations. The JAXA team calculated the continuous natural meteoroid impacts on martian moons. Neither team considered the effect of thermal fatigue, which is likely to occur at a rate orders of magnitude faster than fragmentation due to micrometeoroid impact. The fragmentation of surface material into smaller pieces could significantly enhance the rate at which any biological material present is degraded by exposure to the radiation.

Finding: The committee finds that the sterilizing effect of meteoroid impacts following deposition of martian material on the surface of Phobos and Deimos to be minor due to the low flux of impactors. However, the effects of thermal fatigue could significantly enhance the rate at which any biological matter present is exposed to and degraded by radiation.

TASK 2—RESTRICTED OR UNRESTRICTED EARTH RETURN FOR MARTIAN MOONS SAMPLE RETURN MISSIONS

The second task of the committee was as follows:

> Recommend whether missions returning samples from Phobos and/or Deimos should be classified as "restricted" or "unrestricted" Earth return in the framework of the planetary protection policy maintained by the ICSU Committee on Space Research (COSPAR).

The SterLim team's recommendation on restricted or unrestricted changed between the first and the second meeting of this committee. Initially, their recommendation was for an "unrestricted" mission. However, additional work—conducted between the committee's two meetings—on individual cratering events at Mars ("discrete ejections") in general and the Zunil impact crater in particular, resulted in a recommendation for restricted Earth return. The latter recommendation was based on the uncertainty as to whether an unidentified large (>10 km) young crater might exist on Mars that could have contributed significantly to the deposition of martian materials on Phobos. All such craters less than 20 Ma old appear to have been identified in a search conducted by Werner et al.[2] The SterLim team concluded that unless such large craters can be shown to be ancient (at probabilities over 99.9999 percent), then Phobos cannot be said to be free from hypothetical martian life. They noted that if the Mars ejecta contains low levels of life, the chance of transferring hypothetical martian organisms to Deimos is below the 10^{-6} criterion.

> **Finding:** The committee finds that it is highly unlikely that such a large, young crater exists and has somehow escaped detection.

The JAXA team felt that due to the large uncertainties in many key assumptions used in the models (most importantly, the initial bioload of the martian surface), they were unable to explicitly say whether the probability of sampling a live microbe was greater or less than 10^{-6}. However, they made a comparative argument based on the amount of martian material that arrives on Earth naturally, in the form of martian meteorites, to the amount of martian material expected to be present on Phobos (or Deimos) that would be delivered to Earth by a sample return mission. They investigated another "chain of events," that of ejecta transported from the martian surface to an interplanetary trajectory, and finally hitting Earth, after passing through Earth's atmosphere. Their conclusion was that the natural flux of direct samples from Mars to Earth is orders of magnitude greater that the flux from robotic sample return. Thus, samples returned from the martian moons should be characterized as unrestricted, regardless of the uncertainties.

The JAXA team's argument was based on the inventory of martian meteorites (a total of ~19.2 kg) that have been identified on Earth. This committee has revisited the same argument, to elaborate further this issue.

Ratio of Natural to Spacecraft Flux of Martian Material to Earth

In considering the planetary protection requirements for sample return missions from Phobos and Deimos, it is instructive to compare the natural flux of martian material to Earth relative to that of a robotic sample return mission. Assuming life exists in the surface and near-surface regions of Mars, the flux ratio is a proxy for the relative contribution of spacecraft-transported martian life-forms arriving on Earth compared to those arriving on Earth naturally. The JAXA team calculated that the maximum mixing ratio of martian material to Phobos regolith for the material coming from the Zunil impact was ~100 ppm.[3] Therefore, a 100 g sample collected by a nominal Phobos sample return mission will return 10^{-5} kg of Mars-derived material to Earth. As shall be shown, this mass is negligible compared with the mass of unsterilized Mars material that has arrived on Earth over the last million

[2] S.C. Werner, A. Ody, and F. Poulet, 2014, The source crater of the martian shergottite meteorites, *Science* 343(6177):1343-1346.

[3] See Section 9 of in K. Fujita, K. Kurosawa, H. Genda, R. Hyodo, T. Mikouchi, S. Matsuyama, and the Phobos/Deimos Microbial Assessment Team, 2018, "Assessment of Microbial Contamination Probability for Sample Return from Martian Moons," GNG-2018003, http://sites.nationalacademies.org/SSB/CurrentProjects/SSB_181917.

years since the Zunil impact. Indeed, the argument holds even if there is a comparatively young, large crater that has been overlooked so far.

Following the work of Gladman,[4] a minimum of 15 Mars rocks with a mass of ~100 kg arrive annually.[5] It is known that the preentry masses of Mars rocks may exceed 10 kg.[6] Taking into account only the Zunil impact, the ejected mass is around 2×10^{10} kg.[7] Gladman estimated that ~5 percent of this material eventually will hit Earth and that 0.5 percent hit in the first million years,[8] whereas Mileikowsky et al. estimated that 6 percent hit over a period of 10 million years.[9] The accretion rate of Mars rocks by Earth over the first 10 million years is approximately constant and negligible after that time. Thus, the mass from Zunil that has already hit in the last million years is 10^8 kg. This mass applies to Zunil alone, but within uncertainty (an order of magnitude) it also applies to the flux from the small number of Mars rock producing impacts in the last 10 million years.

Thus, about 10^8 kg of Mars (i.e., 0.05 percent of 2×10^{10} kg) have arrived in the last million years, compared with 10^{-5} kg in the proposed Phobos sample. Therefore, the ratio of spacecraft flux to natural flux over the last million years is 10^{-13}.

The committee compared the relative influx of microbes from the Phobos sample to the natural influx following the series of vicissitudes outlined in the JAXA report (see Chapter 2). The processes that eject material from Mars into space are essentially the same. Rock-size (~kg) pieces are needed to survive transport to Earth and to embed themselves into the regolith of Phobos. Rocks reaching Phobos and Earth both need to be ejected at cosmic velocities. The ejection velocity for Earth is higher than for Phobos. Weakly shocked Mars rocks where microbes would have survived are known,[10] so this effect only modestly reduces the flux to Earth relative to Phobos.

The committee considered sterilization of microbes within Mars rocks on the way to Earth with sterilization of rocks on the surface of Phobos. In both localities, the number of living microbes decreases exponentially with time:

$$N = N_0 \, e^{-t/t_{rad}}$$

where N_0 is the initial number of live microbes, t is time of radiation exposure since leaving Mars, and t_{rad} is the time scale for microbial death.

The JAXA report discusses the Zunil crater on Mars with a rounded age of 1 million years. One in 10^4 buried microbes survived on Phobos, giving t_{rad} = 108,000 years. The most recent impact dominates the microbial load on Phobos. Importantly, any microbe now collected on Phobos has endured this sterilization.

- For Earth-bound ejecta, 1 in 10^8 microbes now arriving at Earth survived radiation, because t_{rad} is 54,000 years. The radiation times differ by a factor of 2 (with Phobos t_{rad} of 108,000 years) because Earth-

[4] B. Gladman, 1997, Destination: Earth. Martian meteorite delivery, *Icarus* 130(2):228-246.

[5] One obtains the total influx of meteorites to the Earth by taking a target of known surface area (for example licensed cars in North America) where impact of a meteorite will be reported and the extrapolating to the full area of the Earth. Then one multiplies by the fraction of Mars rocks ~1/20,000 to total meteorites observed in falls and collections from glacial ice and deserts where there is little sampling bias. Note that Mars rocks resemble Earth rocks. Their provenance is rarely recognized unless they are observed to fall, hit human-made objects, or are collected from deserts and glacial ice. Almost all Mars rocks material fall in the oceans or impact rural areas and are never detected. The tiny fraction of Mars rocks that end up in collections comes into researchers' flux calculations only in determining the ratio of Mars rocks in total meteorites.

[6] Gladman (1997) shows that atmospheric entry results in an ablative mass loss of some 40-to-60 percent. Thus any meteorite with a mass greater than approximately 5 kg had a preentry mass of 10 kg or greater. Ten of the 104 martian meteorites cataloged in "The Martian Meteorite Compendium" (see https://curator.jsc.nasa.gov/antmet/mmc/index.cfm) maintained by Astromaterials Research and Exploration Sciences curation facility at NASA's Johnson Space Center have masses greater than 5 kg. For example, the Tissint and Zagami meteorites have masses of 12 and 18 kg, respectively.

[7] As determined by integrating the curves in Figure 8-7 of the JAXA report above the escape velocity. See Figure 8-7 in Fujita et al., 2018, "Assessment of Microbial Contamination Probability for Sample Return from Martian Moons."

[8] B. Gladman, 1997, Destination: Earth. Martian meteorite delivery, *Icarus* 130(2):228-246.

[9] See p. 399 of C. Mileikowsky, F.A. Cucinotta, J.W. Wilson, B. Gladman, G. Horneck, L. Lindegren, H.J. Melosh, H. Rickman, M. Valtonen, and J.Q. Zheng, 2000, Natural transfer of microbes in space; 1. From Mars to Earth and Earth to Mars, *Icarus* 145:391-427.

[10] B.P. Weiss, J.L. Kirschvink, F.J. Baudenbacher, H. Vali, N.T. Peters, F.A. MacDonald, and J.P. Wikswo, 2000, A lower temperature transfer of ALH 84001 from Mars to Earth, *Science* 290:791-795.

bound martian ejecta receives radiation from all directions, whereas the Phobos sample is irradiated only from above.

- It is important to note that some martian ejecta (meteoroids) reached Earth soon (i.e., months) after the Zunil impact and, thus, essentially escaped radiation sterilization during their journey to Earth.
- The chance of a Mars rock (that eventually reaches Earth) reaching Earth is 1 in 10^{-7} per year and constant at small (greater than a few months and less than few million years) times.[11]
- Thus, about 0.5 percent of the ejecta (that arrived in the last 1 million years since the Zunil impact) arrived during the first t_{rad} and were unscathed by radiation.
- This fraction would dominate the total flux of live microbes. That is, the Phobos fraction depends on a negative exponential of time since the impact, while the Earth fraction of live microbes depends linearly on the decay time.

Next, Phobos organisms needed to survive impact on a solid surface at cosmic velocities. About one in 10^4 microbes are expected to survive. Somewhere between 10 and 100 percent of the microbes in a sample survive passage through Earth's atmosphere. This survival rate comes from the numerical analysis undertaken by the JAXA team. With an atmospheric entry velocity of ~5 km/s, the JAXA team used the same atmospheric sterilization model that it used for Mars and the same sterilization criterion (i.e., heating to 773 K for 0.5 second) to show that meteorites >10 cm across suffer a survival rate of between 20 and 80 percent. This fraction is somewhat misleading, as parts of the Mars rock are strongly heated and sterilized, while other parts of the rock remain unheated and unscathed. Impact at a gentle velocity on Earth's surface is benign. Assuming an overall survival rate of 100 percent for atmospheric entry and landing is, at face value, the conservative choice. However, for the purposes of this calculation, it is preferable to underestimate the number of microbes arriving via martian meteorites. This may appear paradoxical. But the goal is to calculate the ratio of microbes potentially contained in a sample collected on Phobos or Deimos to the natural flux of life-forms to Earth via martian meteorites. Overall, the conservative choice is to maximize the ratio by minimizing the denominator. Therefore, the conservative choice is to assume a 10 percent survival rate during atmospheric entry and landing.

Collecting terms and retaining only orders of magnitude, 10^{-4} of the microbes from Phobos survive impact on its surface. Of these, a fraction of 10^{-4} survived subsequent radiation, giving a net survival of some 10^{-8}.

A fraction of 10^{-2} of microbes evaded radiation by quickly coming to Earth. Of these, a fraction of 10^{-1} reached Earth's surface alive, giving a total survival of 10^{-3}. This ratio of spacecraft to natural live relative microbial fluxes, 10^{-5}, is multiplied by the mass flux ratio of 10^{-13}, to give an overall ratio of 10^{-18}.

Note that this ratio remains small (~10^{-12}) even if the current sterilization factor of 10^{-8} is used for radiation rather than the overall rate of 10^{-2} for rocks not hitting Earth.

For comparison, the committee estimates the flux of live microbes to Earth during the last 50,000 years, when modern humans were present. The mass flux was similar to that in the first 50,000 years after Zunil, 10^8 kg of Mars, compared with 10^{-5} kg in the sample, for a ratio of 10^{13}. Only 1 in 10^8 of the organisms in Mars rocks survived radiation in space, and only 1 in 10 survive entry to the surface of Earth, for a net survival ratio of 10^{-9}. For the Phobos sample, 1 in 10^4 survived impact on Phobos, and of these 1 in 10^4 survive radiation exposure on Phobos, for a net ratio of 10^{-8}. So, the ratio of surviving microbes in Mars rocks to those in the Phobos sample is the mass ratio 10^{13} times the relative survival ratio ($10^{-9}/10^{-8}$), or 10^{12}. Even over the last 50 years, a factor of 10^9 more microbes have arrived in martian meteorites (assuming life actually exists on Mars) from Zunil than will arrive in the sample returned by a spacecraft. Uncertainty in factors, such as the survival probability during Earth entry, will not change the overall conclusion that the mass of naturally delivered martian rocks (and microbes) is many orders of magnitude greater than that delivered by a sample return mission to Mars's moons.

[11] B. Gladman, 1997, Destination: Earth. Martian meteorite delivery, *Icarus* 130(2):228-246.

Restricted Versus Unrestricted Earth Return

In determining whether samples returned from Phobos or Deimos should be classified as restricted or unrestricted Earth return, the committee considered the following factors:

- The work represented by the SterLim and JAXA teams can be considered as the state of the art, in regard to the modeling of the process of deposition of martian material on the surface of the martian moons. Nevertheless, significant deficiencies exist in understanding, and there remain experimental and computational challenges associated with the quantitative estimation of ejecta mass and temperature distributions. Even though issues still exist with the modeling work that was performed (see Chapter 2), the work is convincing in showing that there is significant sterilization introduced during the whole chain of events.
- The issue of desiccation on the surface of the martian moons for any present martian microbes was not considered. At temperatures above the freezing point of water, desiccation is bactericidal to even the most radiation-resistant microbes in a matter of months. (See "Sterilization by Radiation on Phobos/Deimos Surfaces," in Chapter 2.)
- The relative influx of martian microbes from the Phobos/Deimos sample versus the natural influx of direct Mars-to-Earth transfer can be shown to be smaller by several orders of magnitude.

Recommendation: After considering the body of work conducted by the SterLim and JAXA teams, the effect of desiccation on the surfaces of the martian moons, and the relative flux of meteorite- to spacecraft-mediated transfer to Earth, the committee recommends that samples returned from the martian moons be designated unrestricted Earth return.

TASK 3—DIFFERENCES BETWEEN PHOBOS AND DEIMOS IN THE CONTEXT OF PLANETARY PROTECTION

The third task of the committee was to elaborate

> In what specific ways is classification of sample return from Deimos a different case than sample return from Phobos?

The different orbits and cross-sectional areas of Phobos and Deimos result in differences in the velocities associated with impacts of martian ejecta to their surfaces, and also in the total mass of martian material expected to be delivered to each moon. Both of these factors affect the total likelihood that microbes could survive delivery to the moons from Mars, and therefore raises the important question of whether Phobos and Deimos should be treated differently with respect to planetary protection requirements.

The JAXA study concluded that although more martian material was likely to be present on Phobos than on Deimos, more total organisms could theoretically survive transfer from Mars to Deimos. This conclusion was strongly dependent on the specific ejecta geometries and velocities associated with the Zunil impact modeling. In this scenario, martian material impacted Phobos's surface at significantly higher velocities than it impacted Deimos. When coupled with the assumptions about hypervelocity impact sterilization rates as a function of velocity, these impact velocity differences were more significant in determining the probability that microbes could survive transfer than the total mass of delivered martian material.

The committee considered these results in debating the question of whether differences in planetary protection requirements for Phobos versus Deimos are appropriate. The committee uncovered significant uncertainties associated with the impact sterilization assumptions (see "Sterilization During Hypervelocity Impact on Phobos/Deimos Surfaces," in Chapter 2) and noted that choosing a different hypervelocity impact sterilization rate would affect the results from the JAXA work. Specifically, if assumed impact sterilization rates were lowered (especially at high velocities), the effects of differences in impact velocities between Phobos and Deimos may no longer dominate total martian material as the factor that drives total number of microbes that could have survived transfer to Phobos and Deimos.

Recommendation: Given uncertainty associated with impact sterilization assumptions, the committee recommends that Phobos and Deimos should not currently be treated differently in their planetary protection requirements.

TASK 4—RELEVANT INFORMATION FROM STUDIES OF MARTIAN METEORITES

The fourth task was to identify

> What relevant information for classification of sample return is available from published studies of martian meteorites on Earth?

An overview of the literature is included in Chapter 1 (see "Earth Inventory of Martian Meteorites"). The main result from studies of martian meteorites is that coherent, solid rocks may be ejected to Mars's escape velocity (>5.03 km/s). The degree to which the rocks are affected by the impact-ejection process ranges from weakly shocked—represented by igneous clinoproxenites, the nakhlites (5-10 GPa)—to strongly shocked basaltic rocks—the so-called shergottites (~55 GPa).

The current inventory of martian meteorites in collections around the world is biased toward sampling of igneous crust of Amazonian age (i.e., the last 3 billion years of martian history). Despite observation of sedimentary rocks on Mars, no such rocks exist in the known meteorite inventory, suggesting a bias in the delivery process toward young, coherent rocks. However, as no microbe has been so far reported from a martian meteorite, this suggests that the martian bioload is small, possibly smaller than the value assumed by the SterLim and JAXA teams.

Finding: The committee finds that the study of martian meteorites provides important context for studies of Mars and its moons and limited information (e.g., mass and flux to Earth) of relevance to planetary protection considerations. The unambiguous detection of an indigenous martian organism in a meteorite would be of great scientific and societal significance.

TASK 5—PLANETARY PROTECTION CONSEQUENCES OF SAMPLING AT DEPTH

The committee's fifth task was to answer the question

> What are the planetary protection consequences of taking a surface sample at depths of 0-2 cm versus taking a sample extending down to depths of 2-10 cm or deeper?

The most penetrating forms of ionizing radiation suffer little attenuation over the two depth ranges, and so its sterilizing power 2 to 10 cm below the surface is essentially the same as it is in the top 2 cm. However, the committee identified two factors that could cause microbial survival probabilities to be different in these two depth ranges, as follows:

- Ultraviolet radiation, and
- Diurnal temperature cycling.

Ultraviolet radiation would decrease microbe survival rates at the surface of Phobos or Deimos, but such radiation is attenuated within the top few millimeters of surface material. There are therefore unlikely to be significant differences in returned samples due to ultraviolet irradiation from ~0.5-2 cm versus 2-10 cm.

A larger difference for samples from 0-2 cm depth versus the 2-10 cm depth will be the different temperature variations these regions experience. The temperatures on the surface of Phobos and Deimos (see Table 1.1) range, respectively, from ~150 K to 300 K and ~160 K to 270 K according to the time of day.[12,13,14,15] This temperature cycling will add an additional sterilization factor in the upper ~0.3-1.0 cm of Phobos's regolith, but will have little effect at depths below 2 cm because Phobos has a very low thermal conductivity and associated diurnal skin depth.[16] The corresponding surface thermal conductivity of Deimos is assumed to be similar to that of Phobos.

Sterilization by thermal cycling was not considered by the JAXA study, yet the results still showed that acceptably low microbial loads were expected to be present in samples. Therefore, samples from shallower depths on Phobos or Deimos have a lower risk for microbial contamination that those at a greater depth due to sterilization by thermal cycling. However, this additional factor is not needed to give confidence that samples from 2-10 cm depth will be below the established planetary protection limits for expected microbial contamination.

Recommendation: The committee recommends that no differences need to be made in planetary protection requirements for samples collected on the martian moons from depths 0-2 cm versus samples from 2-10 cm.

TASK 6—OTHER REFINEMENTS TO PLANETARY PROTECTION FOR MARTIAN MOONS SAMPLE RETURN

Last, the committee's sixth task was to

> Suggest any other refinements in planetary protection requirements that that might be required to accommodate spacecraft missions to and samples returned from Phobos or Deimos.

The committee limits its response to this task to comments on three specific topics, uncertainty quantification, implications for Mars sample return missions, and the publication of the work of the SterLim and JAXA teams.

Uncertainty Quantification

The work of the SterLim and JAXA teams provides prime examples of attempts to reach a specific conclusion about real-world activities based on combining the results from multiple numerical simulations and laboratory experiments. Each individual calculation or experiment is subject to various degrees of uncertainty such as the following:

- *Parameter uncertainty*—Numbers required as input to models are unknown or poorly constrained—for example, the microbial bioload on the martian surface.
- *Parametric variability*—The range of variables used in modeling may not match the circumstances being modeled—for example, the range of impactor velocities and angles of incidence.
- *Structural uncertainty*—The physics underlying the processes being modeled is not well understood—for example, the choice of the equation of state used in an impact simulation.
- *Algorithmic uncertainty*—Errors or approximations made in the implementation of a particular numerical model need to be understood—for example, the validity of the SPH approach to modeling impacts.

---

[12] R.O. Kuzmin, T.V. Shingareva, and E.V. Zabalueva, 2003, An engineering model for the Phobos surface, *Solar System Research* 37:266, https://doi.org/10.1023/A:1025074114117.

[13] R.O. Kuzmin and E.V. Zabalueva, 2003, The temperature regime of the surface layer of the Phobos regolith in the region of the potential Fobos–Grunt Space Station landing site, *Solar System Research* 37:480.

[14] D.K. Lynch, R.W. Russell, R.J. Rudy, S. Mazuk, C.C. Venturini, H.B. Hammel, M.V. Sykes, R.C. Puetter, and R.B. Perry, 2007, Infrared spectra of Deimos (1-13 µm) and Phobos (3-13 µm), *The Astronomical Journal* 134:4.

[15] J.L. Bandfield, S. Piqueux, T.D. Glotch, K.A. Shirley, T.C. Duxbury, J.R. Hill, C.S. Edwards, J.J. Plaut, V.E. Hamilton, and P.R. Christensen, 2018, "Mars Odyssey THEMIS Observations of Phobos: New Spectral and Thermophysical Measurements," 49th Lunar and Planetary Science Conference, Abstract #2643, https://www.hou.usra.edu/meetings/lpsc2018/pdf/2643.pdf.

[16] R.O. Kuzmin and E.V. Zabalueva, 2003, The temperature regime of the surface layer of the Phobos regolith in the region of the potential Fobos–Grunt Space Station landing site, *Solar System Research* 37:480.

- *Experimental uncertainty*—The observation error and natural scattering of results inherent in any experimental measurement may not be sufficiently well understood absent a sufficient number of repeated measurements—for example, a limited number of impact velocities and angles of incidence explored in laboratory impact studies.
- *Extrapolation and interpolation uncertainty*—The data available from numerical simulations or laboratory studies is insufficient to explore the full parameter range of interest—for example, the use of the results of limited impact experiments to calibrate an analytic function subsequently used to explore a parameter space beyond that explored experimentally.

The quantitative study, characterization, and reduction of uncertainties of these various types is the province of the discipline of uncertainty quantification. Practitioners of uncertainty quantification seek to determine the likelihood of specific outcomes for a system given that specific aspects of it are unknown or only weakly constrained.[17,18] Examples of such uncertainty quantification approaches can be found in other fields—for example, in the Predictive Science Academic Alliance Program (PSAAP) developed by the U.S. Department of Energy,[19] and this area is the subject of a number of conferences and workshops.

Recommendation: The committee recommends that a significant effort be made by the planetary protection community to formally develop an uncertainty quantification protocol that can be used to estimate the cascading uncertainties that result from the integration of multiple computational models or other factors relevant to the quantitative aspects of planetary protection. Specific attention should be given to consideration of the significant uncertainties in the model inputs that exist because of limited available experimental or observational data.

Implications for Mars Sample Return

What, if any, implications do the results of this study and the work of the JAXA and SterLim teams have for the planetary protection aspects of returning samples directly from the surface of Mars? There are several reasons why Mars sample return (MSR) missions differ from those for collecting samples from Phobos and Deimos, including the following:

- The MSR missions will be dedicated to studying the history and evolution of life in four different environments—hydrothermal, sedimentary, subaerial, and rock hosted—according to the recent International MSR Objectives and Samples Team (iMOST) study.[20] The collected samples will be selected according to specific criteria designed to maximize the chance of sampling evidence of extant or extinct life. Therefore, the starting point of the quantitative evaluations undertaken by the SterLim and JAXA teams in terms of the number of colony-forming units in the martian regolith (see "Potential Microbial Density on the Martian Surface," in Chapter 2) will very likely be invalid when considering MSR.
- The various physical processes evoked in the microbial contamination assessment (excavation by impact, collision with Phobos, sterilization, etc.) do not have to be considered for the assessment of potential microbial density for MSR, which could increase drastically the potential microbial density in comparison to the Phobos and Deimos samples.

---

[17] See, for example, R.L. Iman, and J.C. Helton, 1988, An investigation of uncertainty and sensitivity analysis techniques for computer models, *Risk Analysis* 8(1):71-90, doi:10.1111/j.1539-6924.1988.tb01155.x.

[18] See, for example, W.E. Walker, P. Harremoës, J. Rotmans, J.P. van der Sluijs, M.B.A. van Asselt, P. Janssen, and M.P. Krayer von Krauss, 2003, Defining uncertainty: A conceptual basis for uncertainty management in model-based decision support, *Integrated Assessment* 4(1):5-17, doi:10.1076/iaij.4.1.5.16466.

[19] See, for example, P.-H.T. Kamga, B. Li, M. McKerns, L.H. Nguyen, M. Ortiz, H. Owhadi, and T.J. Sullivan, 2014, Optimal uncertainty quantification with model uncertainty and legacy data, *Journal of the Mechanics and Physics of Solids* 72:1-19.

[20] Beaty et al., 2018, "iMOST: Potential Science and Engineering Value of Samples Delivered to Earth by Mars Sample Return," white paper, posted August, by MEPAG at https://mepag.jpl.nasa.gov/reports.cfm.

- The reasoning regarding natural flux does not apply directly to samples returned from the Mars surface. The material will be gently sampled and returned directly to Earth. The sample may well come from an environment that mechanically cannot become a Mars meteorite. The microbes may not be able to survive impact ejection and transport through space. Samples with current liquid water and recent ice seem especially fragile to natural transport to Earth.

Finding: The committee finds that the content of this report and, specifically, the recommendations in it do not apply to future sample return missions from Mars itself.

Publication of the Work of the SterLim and JAXA Teams

Planetary protection policies and the studies underlying them have a reputation in some circles as being based on faulty or outmoded ideas and approaches. The immense amount of work undertaken by the SterLim and JAXA teams makes it clear that these criticisms are, at least in this case, unfounded. Another criticism of planetary protection is a lack of transparency as to how particular conclusions and policies were reached. The planetary protection, astrobiology, and planetary science communities would greatly benefit from the publication of the work undertaken by the SterLim and JAXA teams.

Recommendation: The committee recommends that the SterLim and JAXA teams formally publish the details of and results from their studies or make them readily available in some publicly accessible form.

Appendixes

A

Original Request from NASA

National Aeronautics and Space Administration
Headquarters
Washington, DC 20546-001

MAY 19 2017

Reply to Attn of: Science Mission Directorate

Dr. Fiona Harrison
Chair, Space Studies Board
National Academies of Sciences, Engineering, and Medicine
500 5th Street, NW
Washington, DC 20001

Dear Dr. Harrison:

In accordance with international treaty obligations, NASA maintains a planetary protection policy to avoid biological contamination of other worlds, as well as to avoid the potential for harmful effects on the Earth due to the return of extraterrestrial materials by spaceflight missions. NASA Policy Directive 8020.7 requires that planetary protection requirements be based on recommendations from both internal and external advisory groups, but most notably the Space Studies Board (SSB). NASA relies on the Board's ability to synthesize input from a wide spectrum of the science community and provide expert advice and recommendations, both as an advisory body and as the U.S. representative to the International Council for Science's Committee on Space Research (COSPAR), which is consultative to the United Nations Committee on the Peaceful Uses of Outer Space. As such, the SSB's recommendations on planetary protection are internationally recognized as authoritative and independent of NASA.

Planetary protection requirements for each mission and target body depend upon the type of encounter it will have (e.g., flyby, orbiter, or lander) and the nature of its destination (e.g., a planet, moon, comet, or asteroid). If the target body has the potential to provide clues about the origins and evolution of life or prebiotic chemical evolution, a spacecraft going there must meet a higher level of cleanliness, and some operating restrictions will be imposed. Spacecraft going to target bodies with the potential to support Earth life must undergo stringent cleaning and bioload-reduction processes, and may be subject to greater operating restrictions.

Current COSPAR planetary protection policy does not specify the status of sample return missions from Phobos or Deimos, the moons of Mars. Although the moons themselves are not considered potential habitats for life, recent modeling studies, that were supported by the NASA Planetary Protection Officer in coordination with the ESA Planetary Protection Officer under the ESA-NASA Letter of Agreement on Planetary Protection, indicate that a significant amount of material recently ejected from Mars could be present on the surface of Phobos.

- 2 -

Following this work, ESA issued a contract to perform additional modeling and tests to assess the extent to which post-ejection environmental conditions might inactivate potential Mars life transported to Phobos and Deimos in recent ejecta. The tests included hypervelocity impact inactivation of relevant Earth organisms, as well as ionizing radiation and heat.

Multiple space agencies, including NASA, are developing plans for bringing samples of material from Phobos or Deimos, which need to receive a planetary protection categorization of either Restricted or Unrestricted Earth Return. A designation of Restricted Earth Return, per NASA, ESA, and COSPAR policy, would require samples to be maintained in high containment and undergo a biohazard test protocol after return. In addition, the moons of Mars are targets for future human exploration, so an understanding of the potential for life from Mars to persist is relevant to assuring astronaut safety on those missions.

Furthermore, the European Space Agency (ESA) has ambitious plans for future Mars exploration missions, as does the Japanese Space Agency (JAXA). It is our understanding that ESA has requested the European Science Foundation (ESF) conduct a very similar review of the results of the ESA-funded research, which would include independent Japanese experts supported by JAXA. Given the close working relationship between NASA and ESA, in general, and the Letter of Agreement on Planetary Protection, in particular, the National Academies should engage with ESF and perform a joint study responsive to the needs of both agencies. It is expected that a detailed scope and objectives Terms of Reference for the joint undertaking will be established by consultation between the National Academies and the ESF.

I would like to request that the National Academies establish an ad hoc committee to review the recent ESA-funded work on environments encountered by Mars ejecta during and after the process of deposition on Mars moons, in the context of previous modeling and current understanding of conditions relevant to inactivation of carbon-based life. A Statement of Task is enclosed. The resulting report from the review shall include recommendations for an update of the planetary protection requirements for sample return from Phobos and/or Deimos.

Once agreement on the scope, cost, and schedule for the proposed study has been achieved, the Contracting Officer will issue a task order for implementation. Dr. Catharine A. Conley, Planetary Protection Officer, will be the technical point of contact for this effort and may be reached at cassie.conley@nasa.gov or (202) 358-3912.

Sincerely,

Thomas H. Zurbuchen, Ph.D.
Associate Administrator,
Science Mission Directorate

B

Revised Request from NASA

National Aeronautics and Space Administration
Headquarters
Washington, DC 20546-0001

March 19, 2018

Reply to Attn of: Office of Safety and Mission Assurance

David Smith
National Academies of Sciences, Engineering, and Medicine
500 5th Street NW
Washington, DC 2001

Dear David:

In May 2017, NASA asked the National Academies to establish an ad hoc committee to review and assess recent research sponsored by NASA and ESA, relating to the planetary protection concern that hypothetic Martian life might exist on the surface of the Martian moons, Phobos and Deimos, consequent to their ejection from the surface of Mars following a major impact event.

The Technical Point of Contact for this effort is the Planetary Protection Officer at NASA, which has recently changed from Catherine A. Conley to Lisa M. Pratt, who has reviewed the original statement of task and the timeline to completion. The National Academies' committee has not completed its work and the Japanese Aerospace Exploration Agency (JAXA) has established a second committee to consider whether missions returning samples from Phobos and/or Deimos should be classified as "restricted" or "unrestricted." Consequently, the National Academies committee is asked to address the following three tasks in addition to the original three tasks. In recognition of the likelihood that developing responses to the additional task will require an additional face-to-face committee meeting, NASA requests the final report be submitted by the end of calendar year 2018.

Addendum to statement of task for Martian Moons Study

1. In what specific ways is classification of sample return from Deimos a different case than sample return from Phobos?

2. What relevant information for classification of sample return is available from published studies of Martian meteorites on Earth?

3. What are the planetary protection consequences of taking a surface sample at depths of 0-2 cm versus taking a sample extending down to depths of 2-10 cm or deeper?

Respectfully,

Lisa M. Pratt
Planetary Protection Officer
Office of Safety and Mission Assurance

C

Committee Members and Staff Biographical Information

MEMBERS

DAVID PEARCE, *Chair*, is a professor of environmental microbiology in the Department of Applied Sciences at Northumbria University in the United Kingdom. The underlying theme of his research is the use of microbiology to understand polar ecosystem function and the potential for shifts in biogeochemical activity that may result from environmental change. Dr. Pearce has worked with the British Antarctic Survey as a microbiologist, head of the Genomic Analysis Section of the Biological Sciences Division, and as an aquatic microbial ecologist. His research interests include microbial biodiversity, environmental microbiology, microbial ecology, molecular ecology, microbial physiology, environmental genomics, extremophiles, life in extreme environments, exploring and applying new technology, and the potential of unknown ecosystems. Dr. Pearce is a member of the British Ecological Society and the International Society for Microbial Ecology. He earned his Ph.D. in microbiology from King's College, University of London. Dr. Pearce previously served on the joint National Academies of Sciences, Engineering, and Medicine-European Science Foundation (ESF) Committee on the Review of MEPAG Report on Planetary Protection for Mars Special Regions.

ANDRÉ ANTUNES is a senior lecturer in microbial genetics and an environmental microbiology researcher at Edge Hill University. He also serves as a research consultant at the Computational Bioscience Research Center, at King Abdullah University of Science and Technology. Dr. Antunes researches biological diversity and ecology, with a specialization in microbial responses to deep sea and other extreme environments. He received his Ph.D. in biochemistry at the University of Coimbra in Portugal.

ATHENA COUSTENIS is a director of research with the National Centre for Scientific Research of France and is currently based at the Paris Observatory in Meudon. Dr. Coustenis works in the field of planetology. She earned her Ph.D. in astrophysics and space techniques from the University of Paris. Her research focuses on the use of ground- and space-based observatories to study solar system bodies. Dr. Coustenis's current interests include planetary atmospheres and surfaces, with particular emphasis on the satellites of the giant planets. She is also interested in the characterization of the atmospheres of extrasolar planets. In recent years, she has been leading efforts to define and select future space missions to be undertaken by the European Space Agency (ESA) and its international partners. Dr. Coustenis is the chair of the ESF's European Space Science Committee, of the Committee on Space

Research's Panel for Planetary Protection, and of ESA's Human Spaceflight and Exploration Science Advisory Committee. She has also chaired and served on numerous other ESA and NASA advisory groups.

MICHAEL J. DALY is a professor in the Department of Pathology at the Uniformed Services University of the Health Sciences in Bethesda, Maryland. He is an expert in the study of bacteria belonging to the family Deinococcaceae, which are some of the most radiation-resistant organisms yet discovered. He received his Ph.D. at Queen Mary University of London. His National Academies service includes membership on the Committee on Planetary Protection Standards for Icy Bodies in the Outer Solar System, the Committee on Planetary Protection Requirements for Venus Missions, the Committee on the Origins and Evolution of Life, the Committee on the Astrophysical Context of Life, and the Task Group on the Forward Contamination of Europa.

ABIGAIL A. FRAEMAN is a research scientist in the planetary science section at the Jet Propulsion Laboratory. Dr. Fraeman specializes in the use of infrared spectroscopy to study the surfaces of Mars, Phobos, and Deimos. She is currently a participating scientist on the Mars Science Laboratory Curiosity rover, a co-investigator on the Compact Reconnaissance Imaging Spectrometer for Mars instrument on the Mars Reconnaissance Orbiter, and deputy project scientist for the Mars Exploration Rover mission. Dr. Fraeman received her Ph.D. in Earth and planetary science from Washington University in St. Louis and her B.S. in physics and geology and geophysics from Yale University. She was selected as a participant in the Chinese Academy of Sciences-National Academy of Sciences (NAS) Forum for New Leaders in Space Science in 2015.

ANSGAR GRESHAKE is the curator of meteorite collections for the Museum für Naturkunde in Berlin, where he is also the head of mineralogical preparation facilities. Dr. Greshake studies the formation and classification of martian meteorites, including their phases and metamorphoses, with specific research interest in carbonaceous chondrites. He received his Ph.D. in planetology from Westfälische Wilhelms-Universität Münster.

GUY LIBOUREL is a professor at the Observatoire de la Côte d'Azur in Nice, France. Prior to this, he was at the Centre de Recherches Pétrographiques et Géochimiques in Nancy and an affiliated professor at the Hawaii Institute of Geophysics and Planetology at the University of Hawaii, Honolulu. Dr. Libourel is an experimental cosmochemist whose research focuses on understanding the formation of the first solid in the solar system using high-temperature experimental approaches. His current research is centered on thermal and mechanical properties of the regolith on small solar system bodies. He is a co-investigator on NASA's OSIRIS-Rex and the Japan Aerospace Exploration Agency's Hayabusa 2 asteroid sample return missions. He is also the OSIRIS-REx coordinating scientist for sample analysis for mission sample science in Europe. Dr. Libourel obtained his Ph.D. at the Université Paul Sabatier de Toulouse.

AKIKO M. NAKAMURA is an associate professor in the Department of Planetology at Kobe University, where she performs laboratory impact experiments to study the velocity distribution of fragments from ejecta and the ejecta from particulate layers. These experiments are designed to provide insights into the collisional evolution of small solar system bodies and regolith formation processes. Dr. Nakamura was a co-investigator on the camera system on the Institute of Space and Astronautical Science's Hayabusa I asteroid sample return mission. Dr. Nakamura completed her B.S. in 1988, her M.S. in 1990, and her Ph.D. in 1993, all at Kyoto University.

FRANÇOIS POULET is an astronomer at the Institute of Space Astrophysics (IAS), a joint research unit of the French National Center of Scientific Research (CNRS) and the Université Paris-Sud. Prior to his position at the IAS, Dr. Poulet worked as a research scientist at the NASA Ames Research Center. He studies Mars and small bodies, and his work includes research on the formation and evolution of planetary surfaces. As deputy principal investigator of the MAJIS spectro-imager aboard the JUICE (Jupiter Icy Moons Explorer) mission, Dr. Poulet also has experience in instrumental development. He received his Ph.D. in celestial mechanics and geodesy from the Department of Space Research at the Observatory of Paris.

ROBIN PUTZAR is a senior scientist in the Space Technology Group at the Fraunhofer Institute for High-Speed Dynamics (also known as the Ernst-Mach Institute, or EMI) in Freiburg, Germany. Mr. Putzar has led several large studies investigating the effects of hypervelocity impacts on spacecraft components and geological material, including ballistic limit analyses. His research interests include hypervelocity accelerators, and he has led the design of such accelerators at EMI. Mr. Putzar was a delegate at the Inter-Agency Space Debris Coordination Committee and served on the Program Committee of the European Conference on Space Debris. He is currently chair of the Aeroballistic Range Association. Mr. Putzar has a baccalaureate degree in engineering sciences from Technology University of Berlin.

KALIAT T. RAMESH is a professor in the Department of Mechanical Engineering at Johns Hopkins University. Dr. Ramesh is also director of the Center for Advanced Metallic and Ceramic Systems and director of the Hopkins Extreme Materials Institute. His research interests are in high strain-rate behavior and dynamic failure of materials, nanostructured materials, injury biomechanics, and planetary-scale impact problems. He served as a visiting professor at the University of Cambridge. Dr. Ramesh has published one book, *Nanomaterials: Mechanics and Mechanisms* (2009). After receiving a B.E. from Bangalore University, he continued to Brown University, where he completed his M.S. in engineering. He was awarded his Ph.D. and an additional M.S. in applied mathematics from Brown University. Dr. Ramesh completed a postdoctoral fellowship with the Center of Excellence in Advanced Materials at the University of California, San Diego. Dr. Ramesh previously served on the National Academies Committee on Opportunities in Protection Materials Science and Technology for Future Army Applications.

NORMAN H. SLEEP is a professor of geophysics in the School of Earth, Energy, and Environmental Sciences at Stanford University. Dr. Sleep's research interests include studying convection at the base of the lithosphere and the interaction of the lithosphere with mantle plume material. He is also currently investigating the microphysics of friction and applying the results to nonlinear attenuation and ground damage by strong seismic waves. Dr. Sleep is a fellow of the American Association for the Advancement of Sciences, the Geological Society of America, and the American Geophysical Union. He has received a number of awards for his work, including the James B. Macelwane award, the George P. Woollard Award from the Geological Society of America, and the 2008 Wollaston Medal from the Geological Society of London. Dr. Sleep earned a B.S. in mathematics from Michigan State University and an M.S. and a Ph.D. in geophysics from the Massachusetts Institute of Technology (MIT). He has previously served on the National Academies Committee on Astrobiology and Planetary Science, the Committee on Survey of Surveys: Lessons Learned from the Decadal Survey Process, the Committee on Earth Resources, and the Committee on Planetary Biology and Chemical Evolution, and he currently serves as the NAS Section 15 liaison.

SHINO SUZUKI is a senior researcher at the Japan Agency for Marine Earth Science and Technology at the Kochi Institute for Core Sample Research in Japan. Dr. Suzuki has researched and published dozens of papers on microbial communities. Previously, she worked at J. Craig Venter Institute, where she collaborated with the NASA Astrobiology Institute developing and employing field, laboratory, and genomic modeling approaches aimed at detecting and characterizing subsurface microbial life. Dr. Suzuki earned her Ph.D. in molecular microbiology from the University of Tokyo.

MEGAN BRUCK SYAL is a physicist in the Design Physics Division at Lawrence Livermore National Laboratory (LLNL). Dr. Bruck Syal specializes in experimental and numerical simulation of planetary impacts, including hypervelocity impact experiments (with an emphasis on porous and volatile-rich materials) and modeling of impact events in a variety of shock physics codes. Her published and ongoing research includes impact delivery of carbon and volatiles to Mercury and the Moon, the excavation of Stickney Crater at Phobos, analysis of impactor- and target-derived vapor plumes using high-speed emission spectroscopy, and giant-impact formation of moons in exoplanetary systems. Additionally, Dr. Bruck Syal is very active in the field of planetary defense, supporting NASA's DART (Double Asteroid Redirection Test) mission with simulations of the planned 2022 spacecraft impact at Didymos B, NASA-Federal Emergency Management Agency Asteroid Impact Tabletop Exercises, and a NASA-National Nuclear Security Administration interagency collaboration on hazardous asteroid mitigation case studies. Her planetary defense work focuses on numerical simulation of deflection and disruption techniques, with a particular emphasis on understanding sensitivities to asteroid initial conditions. Previously, Dr. Bruck Syal was a postdoctoral researcher at LLNL (2014-2016), a Ph.D. candidate in the Geosciences Department at Brown University (2009-2014), and a data specialist at the Smithsonian Astrophysical Observatory's Chandra X-Ray Center (2007-2009). She is a recipient of a NASA Earth and Space Science Fellowship, a NASA Group Achievement Award (Deep Impact-EPOXI mission Science Team), and a Brown University Graduate Fellowship. Dr. Bruck Syal obtained her Ph.D. in planetary geosciences at Brown University.

ERIN L. WALTON is an associate professor at MacEwan University. Dr. Walton holds a Discovery Grant awarded by the Natural Science and Engineering Research Council of Canada. The focus of her research is shock metamorphism of astromaterials, with an emphasis on martian meteorites. Dr. Walton's research interests also encompass the age and formation of terrestrial impact structures, such as the 91-million-year-old Steen River Impact Crater in Alberta. She earned her Ph.D. in geology from the University of New Brunswick.

STAFF

EMMANOUIL DETSIS has been a science officer at ESF since 2012. Dr. Detsis's main role at ESF consists of project management, business development, and coordination for the European Framework program activities of ESF. He has also managed several projects sponsored by the European Space Agency and the European Commission. Dr. Detsis received a B.Sc. in physics from the University of Crete and a Ph.D. in astrophysics from the University of Edinburgh. He also holds a master's degree in space science from the International Space University in France.

DAVID H. SMITH joined the Space Studies Board (SSB) as a senior staff officer in 1991. Dr. Smith has been and is the study director for numerous National Academies' activities in the general areas of astrobiology, planetary science, and planetary protection. He also organizes the SSB's Lloyd V. Berkner Space Policy Internships and the joint NAS-Chinese Academy of Sciences Forum for New Leaders in Space Science. Dr. Smith received a B.Sc. in mathematical physics from the University of Liverpool in 1976, achieved the honors standard in Part III of the Mathematics Tripos at the University of Cambridge in 1977, and received a D.Phil. in theoretical astrophysics from Sussex University in 1981. Following a postdoctoral fellowship at Queen Mary College, University of London, he held the position of associate editor and, later, technical editor of *Sky and Telescope*. Immediately prior to joining the staff of the SSB, Dr. Smith was a Knight Science Journalism Fellow at MIT.

MIA BROWN joined the SSB as a research associate in 2016. Ms. Brown comes to the SSB with experience in both the civil and military space sectors and has primarily focused on policies surrounding U.S. space programs in the international sector. Some of these organizations include NASA's Office of International and Interagency Relations, Arianespace, the United Nations Office for Disarmament Affairs (Austria), and the Department of State. From 2014 to 2015, Ms. Brown was the managing editor of the *International Affairs Review*. She received her M.A. in international space policy from the Space Policy Institute at the Elliott School of International Affairs. Prior to entering the Space Policy Institute, Ms. Brown received her M.A. in historical studies from the University of Maryland, Baltimore County, where she concentrated in the history of science, technology, and medicine and defended a thesis on the development of the 1967 Outer Space Treaty.

ANDREA REBHOLZ joined the Aeronautics and Space Engineering Board (ASEB) as a program coordinator in January 2009. Ms. Rebholz began her career at the National Academies in October 2005 as a senior program assistant for the National Academy of Medicine (formerly the Institute of Medicine) Forum on Drug Discovery, Development, and Translation. Prior to the National Academies, she worked in the communications department of a DC-based think tank. Ms. Rebholz has a B.A. in integrative studies—event management from George Mason University's New Century College. She earned the certified meeting professional designation in 2012 and the certified government meeting professional designation in 2017. She has more than 15 years of experience in event planning, project administration, and editing.

JONATHAN LUTZ is an undergraduate student at the University of Colorado, Boulder, where he is seeking a degree in astrophysics. He interned with the SSB in fall 2018. Mr. Lutz worked as a student associate at the Laboratory for Atmospheric and Space Physics in Boulder, Colorado. He was a member of a student-led BalloonSat research team that launched a scintillator gamma-ray detector on a small payload to the stratosphere. Previously, he worked as a freelance graphic designer and has a background in data science.